Concepts in Chemistry

Structure and Behaviour

SI units

Concepts in Chemistry

Titles in this series:

An Introduction to Chromatography
Critical Readings in Chemistry
A Guide to Understanding Basic Organic Reactions
Orbitals and Chemical Bonding
An Introduction to Chemical Energetics
Spectroscopy in Chemistry
The Chemical Electron
Crystals and Crystal Structure
An Experimental Introduction to Reaction Kinetics
Organometallic Chemistry of the Main Group Elements
Structure and Behaviour

Structure and Behaviour

SI units

Kenneth Wild BSc FRIC

Formerly Head of Science Department, Bromsgrove County High School

and

John Nellist BSc

Assistant Chemistry Master, Bromsgrove County High School

Longman

Longman Group Limited
London
Associated companies, branches and representatives throughout the world

First published 1970
Second impression 1973

ISBN 0 582 32148 4

Printed in Hong Kong by
Yu Luen Offset Printing Factory Ltd

Preface

This book is intended as an introduction to the structure of inorganic substances and in particular to the relationship between structure and behaviour. We have tried to show that structure is an important underlying principle; this principle is kept continually before the student.

The book is written for students working for Advanced Level of the General Certificate of Education and for University Scholarship examinations.

In spite of the current accent on teaching for sense, school inorganic chemistry can easily degenerate into a catalogue of substances, their sources, preparations, properties, and uses only vaguely linked to underlying principles. Too often specially chosen specific properties are used to support a proposed structure rather than to consider the structure as a conceptual model to be used by the student to explain behaviour; nor is the average sixth-former encouraged by the too early introduction of structures based upon sophisticated premises or demanding mathematical treatment.

We have assembled structures based upon electron transfer or electron sharing between atoms for most of the substances encountered in sixth-form inorganic chemistry. A wide range of substances has been grouped into relatively few structural types. Chapter 10 is an attempt to arrange these substances and types in a tabular form.

The student (or the authors for that matter) may not always be correct in the structure proposed or the interpretation given, but we feel that it is better to have tried and failed than never to have tried at all!

A number of books which we have found useful are listed at the end and we acknowledge our debt to the authors concerned.

K.W.
J.N.
Bromsgrove

Contents

Introduction

Why is oxygen a gas? Why is sulphur a yellow solid? Why is nitrogen inert? Why does phosphorus exist in several forms? Why is iron magnetic? How can two substances as different as diamond and graphite be made up from the same element? Why are sodium chloride crystals cubic? What are complex ions? Why is calcium chloride readily soluble in water but calcium sulphate only slightly soluble? Why is a solution of hydrogen chloride in water acidic when a solution of the same gas in toluene is not? What is a covalent bond? Why are the noble gases so called? Why is aluminium chloride vapour dimerised? Why is oxygen paramagnetic? Why do sodium nitrate and lead nitrate behave differently when heated? What happens when ice melts? Why is pure iron malleable and yet iron with a small percentage of carbon brittle? Why do we use conceptual models in science? Why do metals conduct electricity? Why when molten sulphur is heated does the viscosity increase sharply and then decrease? Why is $CaCl_2$ stable while CaCl and $CaCl_3$ do not exist?

Why, why, why?

You may have asked questions like these and many others in your study of chemistry. You may have felt bewildered, even dismayed, by the wide variety of apparently unrelated facts and exceptions to empirical rules which you have encountered. Clearly the need is for unifying ideas, for theories, which serve to bind the subject together and are an aid to understanding.

The periodic table is an important milestone in the search for order. It correlates an abundance of experimental data, reveals patterns of behaviour and allows predictions to be made. It does not, however, provide explanations for the degree of order which it shows to exist. Indeed, it raises further questions rather than suggesting answers.

Where are we to look for concepts which we can use to explain properties and to provide a framework for such diversity? A consideration of models for the structures of substances and the relationship between the proposed structure and the observed properties is an obvious choice; a study of energy changes in chemical reactions also seems promising. In this book we intend to concern ourselves mainly with the structural point of view and have outlined structural models for a number of substances which are commonly met in the laboratory; for many of them an attempt is made to relate the suggested structure to the properties. However, in trying to seek reasons for behaviour it is neither possible nor desirable to divorce a structtural interpretation from other considerations, such as energy changes, and these factors receive some attention.

When we discuss behaviour in terms of atoms we use models, for instance the planetary model of an atom with its central nucleus and orbiting electrons. In this book, the conceptual model used, in an attempt to interpret covalent structures, is termed the *valence-bond model.* The formation of bonds, considered from the point of view of individual atoms, involves the pairing of electrons with opposed spins and the maximum overlapping of atomic orbitals containing these electrons. This model has been chosen in preference to others because it lends itself readily to the problems we are likely to encounter at this stage without involving unnecessary complexity.

The valence-bond model, like any other, has limitations and there are instances where it is not possible to explain experimental observations by its use. These failures do not, however, detract from its general usefulness, and serve to remind us that we are employing a model to interpret observations and are not stating absolute truths.

Some of the difficulties encountered can be resolved by modifying the model or by using other approaches such as the concept of resonance or the molecular orbital theory. These, though more sophisticated, have their limitations and we have chosen not to deal with them but would anticipate that you might well wish to follow them up in alternative texts.

A certain amount of background knowledge is assumed including an acquaintance with the electron configurations of the elements, simple atomic orbital theory and such concepts as van der Waals' forces. The books in the reading list will be helpful.

We began this chapter by asking a number of questions to whet your appetite and to hint at the range and complexity of inorganic chemistry. A consideration of the structure of substances might suggest answers and provide some rationalization of the wide range of properties which we observe. The chapters which follow will go some way towards this goal although it is likely that they will raise some questions while answering others.

Structural interpretation of the behaviour of substances is a topic in which speculation and theorizing can be rewarding and exciting. It gives opportunities for a critical look at the structures, suggested in this and other books, and the chance to propose structures of your own to explain the properties of any of the substances which you encounter.

1 Noble Gases

The only substances to exist as free atoms at room temperature and pressure are the gases—helium, neon, argon, krypton, xenon, and radon. The atoms of all other elements engage in some form of electron rearrangement with each other to produce a variety of structural types. These range from simple diatomic molecules to more complex polyatomic ones and giant structures. The familiar gases hydrogen, chlorine, oxygen, and nitrogen are well-known examples of the diatomic class, and may be generally represented as X_2.

The question arises, why should the six noble gases, and no other elements exist in the monatomic state at room temperature and pressure? Putting this another way, if we ask why do the remaining ninety odd elements enter into electron rearrangements, we answer by saying that they do so to attain greater stability. It follows that the noble gases possess electron configurations which are inherently stable and which cannot readily be improved upon by fresh arrangements of the electrons.

1.1 Electron configurations of the noble gases

Table 1

He	K_2					
	$1s^2$					
Ne	K_2	L_8				
	$1s^2$	$2s^22p^6$				
Ar	K_2	L_8	M_8			
	$1s^2$	$2s^22p^6$	$3s^23p^6$			
Kr	K_2	L_8	M_{18}	N_8		
	$1s^2$	$2s^22p^6$	$3s^23p^63d^{10}$	$4s^24p^6$		
Xe	K_2	L_8	M_{18}	N_{18}	O_8	
	$1s^2$	$2s^22p^6$	$3s^23p^63d^{10}$	$4s^24p^64d^{10}$	$5s^25p^6$	
Rn	K_2	L_8	M_{18}	N_{32}	O_{18}	P_8
	$1s^2$	$2s^22p^6$	$3s^23p^63d^{10}$	$4s^24p^64d^{10}4f^{14}$	$5s^25p^65d^{10}$	$6s^26p^6$

It can be seen from Table 1 that five of the gases have eight electrons (s^2p^6) in the outer, valency shell. Helium is exceptional with two electrons in the K shell (s^2). That these electron configurations are unusually stable can be inferred from the energy level diagram (Fig. 1) and from the ionization energies of the noble gases (Table 2).

1.2 Energy level diagram

The circles on the energy level diagram (Fig. 1) represent *orbitals*, each of which can hold two electrons, which are supposed to be spinning in opposite directions. According to the *aufbau principle* each orbital is filled in turn from the lowest energy level upwards. The diagram shows a large break in energy at each of the *p* levels, and the noble gases are unique in having structures in which all the orbitals up to, and including, the *p* level are filled. In consequence, no orbital is unoccupied and no orbital contains an electron of unpaired spin. For example, krypton, with a configuration of $1s^2 2s^2 2p^6 3s^2 3p^6 3d^{10} 4s^2 4p^6$, has all the orbitals fully occupied.

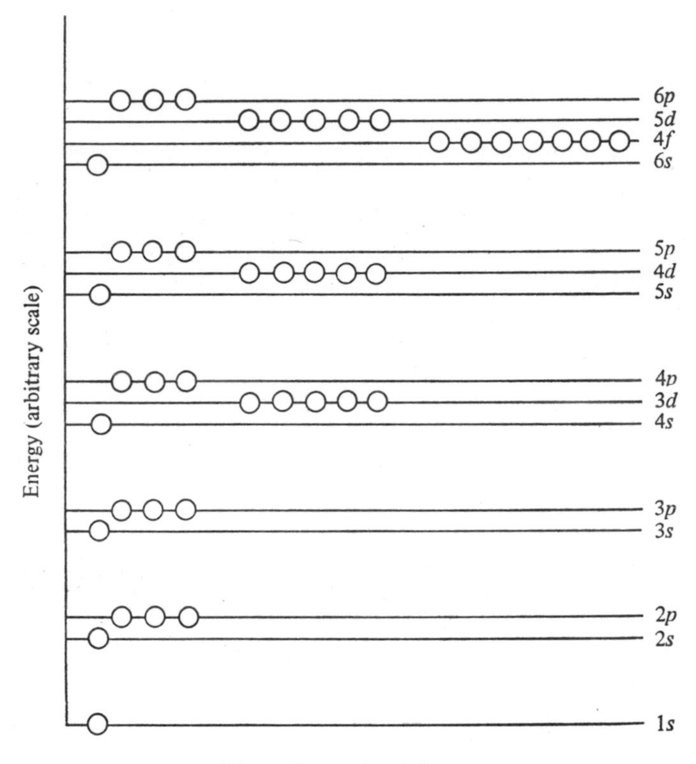

Fig. 1. Energy level diagram

Since the two electrons which fill one orbital have opposed spins, the resultant spin momentum is zero and interaction with electrons in other atoms seems unlikely. From this viewpoint, it is not surprising that the noble gases are monatomic and relatively inert. The energy difference between any *p* level and the immediately higher level is considerably greater than the difference between the same *p* level and the one immediately below.

1.3 Ionization energy

The amount of energy required to remove the most loosely bound electron from a gaseous atom is called the *first ionization energy*. The values for the noble gases are high compared to those for neighbouring elements in the periodic table. Neon is flanked by fluorine and sodium for which the ionization energies are 1682 and 494 kJ mol^{-1} of F and Na respectively. Neon is less likely than sodium or fluorine to form a positive ion by loss of an electron.

Table 2. First ionization energies of noble gases

Symbol	*First ionization energy* (kJ mol^{-1})
He	2372
Ne	2079
Ar	1519
Kr	1347
Xe	1163
Rn	1033

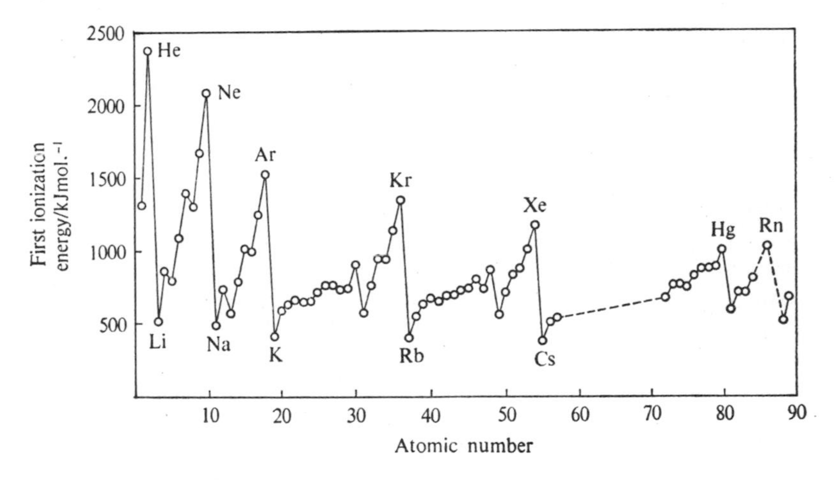

Fig. 2. Relationship between first ionization energy and atomic number

The way in which ionization energy and electron configuration are related can be inferred from Fig. 2 in which ionization energy is plotted against atomic number. The noble gases occupy peak positions indicative of their unreactivity. The alkali metals, on the other hand, are in the troughs, suggestive of the relative ease with which electrons can be removed to form ions. Alkaline earths and transition metals occupy relatively low positions and are correspondingly electropositive. In contrast, the halogens, with little tendency to form positive ions, are on ascending slopes not far below the noble gases.

The periodic relationship between ionization energy and electron configuration is a striking feature of the graph and is a special case of the kind of curve first conceived by Lothar Meyer in 1869.

1.4 Physical properties of noble gases

Table 3. Physical properties of noble gases

Property	He	Ne	Ar	Kr	Xe	Rn
Relative atomic mass	4·0026	20·179	39·948	83·80	131·30	222
Melting point (K)	3	24	84	116	161	202
Boiling point (K)	4	27	87	121	165	211
Solubility in water $cm^3\ dm^{-3}$ at 293 K	13·8	14·7	37·9	73	111	215
Density $g\ dm^{-3}$ at 273 K	0·18	0·90	1·78	3·71	5·85	9·73
Heat of fusion ($kJ\ mol^{-1}$)	—	0·34	1·13	1·51	2·05	3·35
Heat of vaporization ($kJ\ mol^{-1}$)	0·08	1·84	6·28	9·67	13·68	16·40

The physical properties of the noble gases vary, from helium to radon, in a fairly regular manner. The degree of regularity is not matched by any other group of elements in the periodic table. Further, the noble gases approach ideality, either as gases or in solution, more closely than any other substance. These features are the direct consequence of their symmetrical structures.

Look at the water solubility data given in Table 3. Low values are to be expected since distortion of an inert gas atom by a water molecule is slight, and the forces of attraction between water dipoles are much greater than the van der Waals' forces between noble gas atoms. Hence there is little to favour the dissolution of the gases in water. The change in solubility with atomic number can be qualitatively explained in the following way. As the nuclear charge increases so does the screening effect of the inner electrons leaving the outermost electrons more vulnerable to the attractions of the water dipoles, resulting in an increase in water solubility from helium to radon.

1.5 Chemical properties of noble gases

The chemical inertness of the noble gases was until 1962 almost taken for granted. It provided, and to some extent still does, the key to an understanding of chemical reactivity. The electron configuration s^2p^6, the '*stable octet*', became the hallmark of stability and provided the basis for the development of models for the ionic bond and the covalent bond.

The discovery of xenon hexafluoroplatinate ($XePtF_6$) by Bartlett in 1962 paved the way for further preparations of noble gas compounds, and showed the fallibility of assuming a unique category of 'inertness' for the 'octet' without invalidating, as we shall see, the theory for many substances.

Despite the common acceptance that the noble gas configuration is an unusually stable one, there was a good deal of speculation prior to 1962 that chemical reaction involving noble gases ought to be possible, especially for those of higher atomic number. The first ionization energy of radon (1033 kJ mol^{-1}) is little in excess of that of mercury (1004 kJ mol^{-1}). Again the first ionization energy of xenon (1163 kJ mol^{-1}) is less than that of hydrogen (1310 kJ mol^{-1}) or chlorine (1255 kJ mol^{-1}) and yet the hydrogen ion, H^+, and the chlorine ion, Cl^+, are known species, even though they are ordinarily coordinated to a solvent molecule or some other base.

All the noble gases except helium, have four lone pairs of electrons, and since water, ammonia, ether, and many other compounds form coordinate bonds with electron acceptors, it would not be unexpected if the noble gases behaved similarly. Chemical combination by extension of the valency shell beyond eight electrons is common among elements from the second period onwards, and the possibility of this mode of behaviour for noble gases cannot be ruled out.

The lack of success, not for the want of trying for so many years, was the result of attempting to prepare compounds at low temperatures on the assumption, false as it turned out, that they would have no thermal stability.

2 Covalent Molecules

2.1 Hydrogen

The hydrogen atom, consisting of a one proton nucleus and one orbiting electron, is the simplest of all atoms. The nucleus is unshielded by any inner electrons and the electron which occupies the 1*s* orbital is unpaired.

A *covalent bond* is the result of two unpaired electrons on separate atoms being paired and shared; the orbitals are imagined to overlap. Two hydrogen atoms, by sharing and pairing two electrons, one from each atom, fill the overlapped 1*s* orbitals and produce in effect, a helium configuration (Fig. 3).

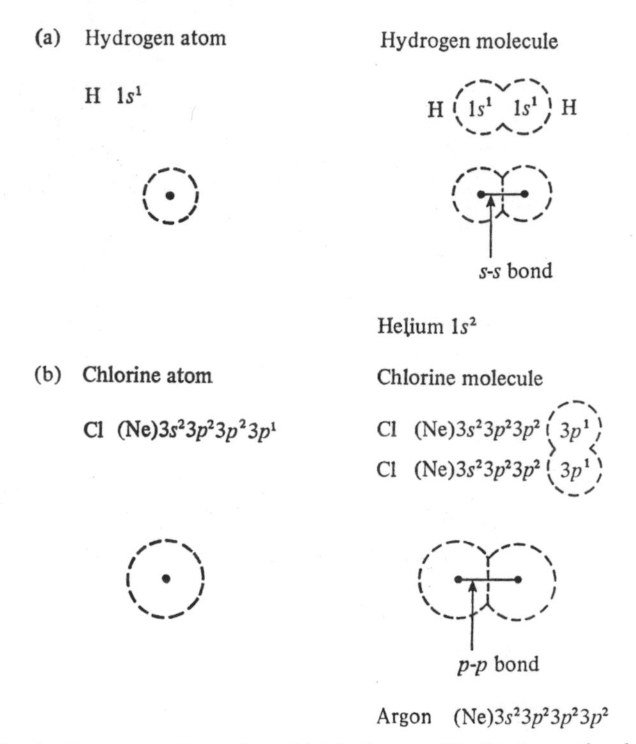

Fig. 3. Electron configuration of (a) hydrogen, (b) chlorine molecules

As a result of the overlap an electron originally associated with one atom exerts a resultant force of attraction for the nucleus of the other atom; hence producing a bond.

It may be expected that the hydrogen molecule is more stable than the hydrogen atom, and this prediction is supported by the facts, since hydrogen gas is diatomic and relatively stable. The strength of the covalent bond joining two hydrogen atoms, corresponds to the lowering of energy resulting from the orbital overlap when the nucleus of one atom attracts the electron of the other.

Since a noble gas configuration is achieved by this sharing process, it might be assumed that the hydrogen molecule is stable, in the sense that a noble gas is stable, and that hydrogen gas might behave in a similar fashion. In terms of physical properties, this prediction is fairly well borne out. The low melting point, boiling point and water solubility suggest that the forces between hydrogen molecules are small as might be expected for a symmetrical molecule with a helium-like electron structure. However, hydrogen is not at all like a noble gas in its chemical properties; it burns readily in air or chlorine and forms other hydrides with metals and non-metals. Can this behaviour be reconciled with the suggested structure?

The explanation is that the bond between two hydrogen atoms can be distorted, even broken, by sufficiently energetic conditions, and in the presence of other substances new electron arrangements may form with possibly greater stability than the hydrogen molecule. This is simply illustrated by the formation of hydrogen chloride. The electron rearrangement which takes place in this reaction after its initiation by heat or ultraviolet light, produces a more stable system than the original chlorine–hydrogen system from which it arose; the reaction is consequently exothermic.

$$\tfrac{1}{2}H_{2(g)} + \tfrac{1}{2}Cl_{2(g)} \rightarrow HCl_{(g)} \qquad \Delta H = -92 \text{ kJ mol}^{-1}$$

2.2 Chlorine

Chlorine, like hydrogen, is diatomic. Its atom is one electron short of a noble gas configuration and has an unpaired electron. Not surprisingly, the covalence mechanism suggested for hydrogen, leads to a structure for the chlorine molecule (Fig. 3). The *p–p* orbital overlap pairs off the unpaired electron on each chlorine atom to give a more stable arrangement suggestive of the structure of argon.

As in the case of hydrogen we must be careful not to assume too great a lack of reactivity from the model for chlorine, otherwise it will be at variance with the chemistry of the substance. Reactivity is relative, environment and conditions must be stated and although the chlorine molecule is less reactive than the lone atom, the electron cloud constituting the covalent bond can be distorted and broken. In the presence of atoms of another element, new and perhaps more stable arrangements may form as was indicated in the formation of hydrogen chloride.

One major difference between the structures of hydrogen and chlorine is that the chlorine molecule has unshared pairs of electrons which could form new bonds either permanently or temporarily with other atoms; such bonds are termed *coordinate bonds* and are considered in greater detail in Chapter 5. The formation of permanent bonds by the use of these unshared pairs is illustrated when the structure of aluminium chloride is dealt with (Fig. 21). It seems reasonable that the temporary formation of bonds using these electrons might well be a step in some chemical reactions which ultimately produce a new and more stable electron arrangement. Such reactions might therefore take place more readily than would be predicted by simply considering the energy required to break the original bonds. In other words a suitable reaction pathway might be available.

A further significant difference between chlorine and hydrogen is that the former could expand its octet beyond the requirements of a noble gas configuration. This may lead to the formation of temporary or permanent bonds through the acceptance of a pair of electrons from a suitable donor atom or molecule, again offering a possible reaction pathway to a more stable electron arrangement. This may be another factor in explaining the discrepancy between the behaviour of chlorine, predicted on the basis of the simple noble gas model and its chemical reactivity.

By considerations similar to those outlined above the other halogens can be expected to be diatomic and to have structures and properties similar to those of chlorine. This expected similarity is on the whole well borne out with one or two notable exceptions.

Some of the more important physical properties of the halogens are listed in Table 4.

The simple molecular structure common to the halogen family with the resulting weak van der Waals' forces operating as attractions between molecules accounts for the high volatility and relatively low heats of fusion and vaporization. The variation of these properties within the family is qualitatively explained by similar reasoning to that used for the noble gases. The magnitude of van der Waals' forces is proportional to the population of electrons and inversely proportional to the tightness with which they are held. Hence, the attractions between the molecules are weakest in fluorine and strongest in iodine; a point well illustrated by the observation that fluorine is a gas and iodine a solid at room temperature.

The colours of the halogen elements can also be correlated to some extent with the structural model of the molecules. The colours arise from absorption of visible light by the molecules, which results in the excitation of one or more of the outer electrons to higher energy levels. It would seem that the energies required for these molecular excitations follow the same pattern as the trend in the ionization energies, which indicate the energy needed

Table 4. Physical properties of halogens

Property	F	Cl	Br	I
Relative atomic mass	18·9984	35·453	79·904	126·9044
Configuration	(He)$2s^2 2p^2 2p^2 2p^1$	(Ne)$3s^2 3p^2 3p^2 3p^1$	(Ar)$4s^2 4p^2 4p^2 4p^1$	(Kr)$5s^2 5p^2 5p^2 5p^1$
Colour	Pale yellow	Green	Dark red	Violet
Melting point (K)	53	171	266	387
Boiling point (K)	85	238	331	456
Heat of fusion (kJ mol^{-1})	0·80	3·22	5·27	7·87
Heat of vaporization (kJ mol^{-1})	3·14	10·21	15·27	20·84
Heat of dissociation (kJ mol^{-1})	154·8	242·7	192·5	150·6

to remove an electron from a free atom in the gas phase. Fluorine, which has the highest ionization energy, absorbs high energy radiations from the violet end of the visible spectrum, and so appears yellow, whereas iodine, with the lowest ionization energy, absorbs yellow and green radiation of lower energy and so appears violet.

The variation in chemical properties of the halogens cannot be explained quite as readily since it requires the consideration of a series of interdependent factors, including the relative sizes of halogen atoms, their electronegativities, their ability to accept and donate electron pairs, and the strengths of the halogen–halogen and halogen–other element bonds. Some of these factors are considered in Chapter 3. However, there is a general similarity in the types of reactions in which the halogens participate, and this can be related to their comparable electron configurations.

2.3 Oxygen

Oxygen is also diatomic but the electron configuration of the oxygen atom shows two unpaired electrons and it is two electrons short of a noble gas configuration. Two electrons on each atom pair giving two covalent bonds between the atoms, and a neon-like configuration (Fig. 4). This model of the oxygen molecule explains the behaviour of the gas in most respects, but we shall return to the structure of oxygen in Chapter 9. Chemical reactivity, as before, is the consequence of fresh electron arrangements in a favourable environment, and physical properties are characteristic of a simple molecular structure.

2.4 Nitrogen

Nitrogen is diatomic, and the nitrogen atom shows three unpaired electrons; the atom is three electrons short of a neon configuration. Logically the molecule should have three covalent bonds, formed by pairing and sharing each unpaired electron between the two nitrogen atoms (Fig. 4).

The characteristic unreactivity of nitrogen is not surprising when the shortness of the nitrogen–nitrogen bonds is taken into consideration, and the fact that the three $2p$ orbitals mutually at right angles, on each nitrogen atom, overlap to give strain-free bonds. The strength of this nitrogen–nitrogen triple bond is borne out by measurement of the heat of dissociation of the nitrogen molecule. It is found to require 941 kJ to split a mole of nitrogen into atoms. In fact nitrogen is the most stable diatomic molecule. The triple bond shows no tendency under ordinary conditions to undergo addition reactions such as are observed for acetylene which also contains a triple bond, and the high heat of dissociation of nitrogen suggests that it is only likely to react at high temperatures, as when magnesium burns in the gas.

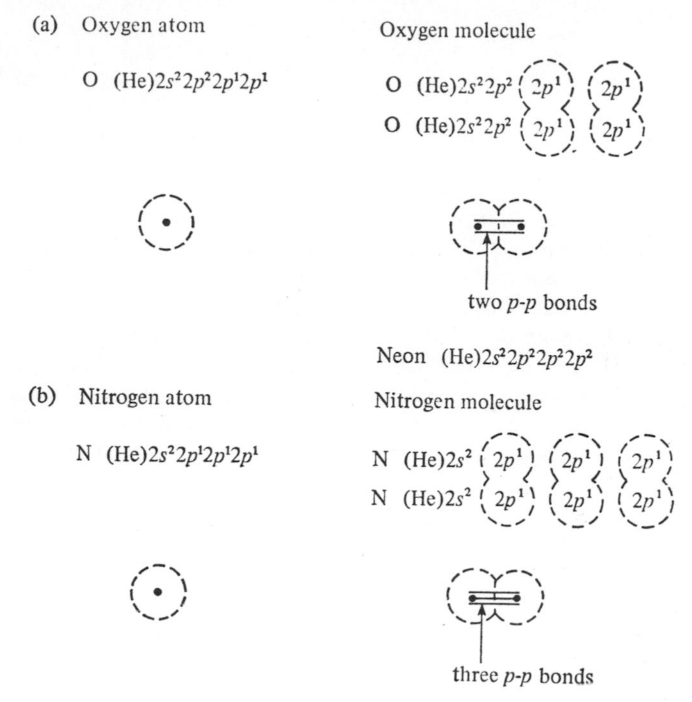

Fig. 4. Electron configuration of (a) oxygen, (b) nitrogen molecules

2.5 Phosphorus

The electron configuration of the phosphorus atom is (Ne)$3s^2 3p^1 3p^1 3p^1$ and the three unpaired electrons suggest a covalency of three in the molecule. Diatomic molecules similar to nitrogen might be expected, but this is not the case below 1873 K, presumably because the phosphorus nuclei would be too far apart to give sufficient overlap of the orbitals for a stable molecule. Instead two, or possibly more, varieties of phosphorus exist. Two of these, white phosphorus and red phosphorus (Table 5), are commonly met in the laboratory.

The red allotrope is less volatile and less reactive suggesting a greater molecular complexity and larger cohesive forces than the white variety.

Table 5. Properties of white and red allotropes of phosphorus

Property	*White Phosphorus*	*Red Phosphorus*
Melting point (K)	317	863
Boiling point (K)	553	1704 (sublimes)
Solubility in carbon disulphide	Soluble	Insoluble
Toxicity	Toxic	Non-toxic
Air	Spontaneous ignition at 303 K	Ignites at 533 K
Alkali	Readily reacts. Phosphine produced	No reaction

Molecular weight determination for the white allotrope shows the existence of the P_4 molecule. This can be explained by assuming a tetrahedral arrangement of phosphorus atoms where each atom engages in *p–p* overlap with three neighbours (Fig. 5). Only weak van der Waals' forces operate between the P_4 molecules which are thus readily separated. Hence, volatility, solubility in covalent solvents, and enhanced reactivity would be expected. Each phosphorus atom in the molecule has an unshared pair of electrons which can, in contrast to nitrogen, enter into the formation of new bonds, permanent or temporary; again offering a possible reaction route in a suitable environment.

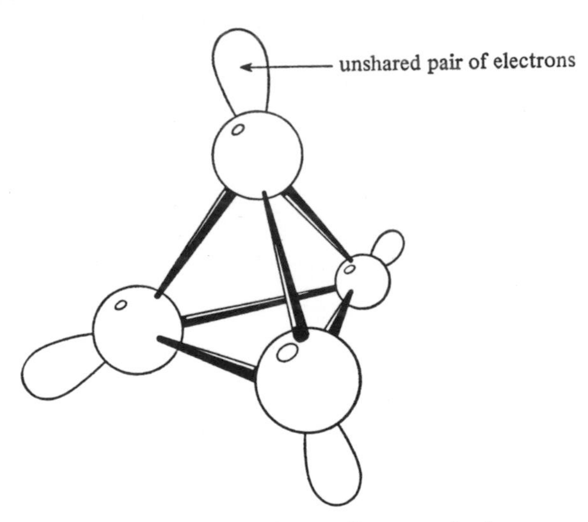

Fig. 5. White phosphorus, discrete molecules

A further factor contributing to the reactivity of white phosphorus is that the *p–p* bonds in the tetrahedral P_4 units are under a certain amount of strain because of the distorting effect of the unshared pairs of electrons. For instance, in the reaction of white phosphorus with alkali, preliminary breaking of a *p–p* bond is likely to be the first and most important step.

It is possible that red phosphorus consists of puckered sheets of phosphorus atoms with a network of covalent bonds. The three 3*p* orbitals on each phosphorus atom are mutually at right angles, and the *p–p* overlap of each phosphorus atom with three neighbouring atoms gives a puckered sheet structure (Fig. 6).

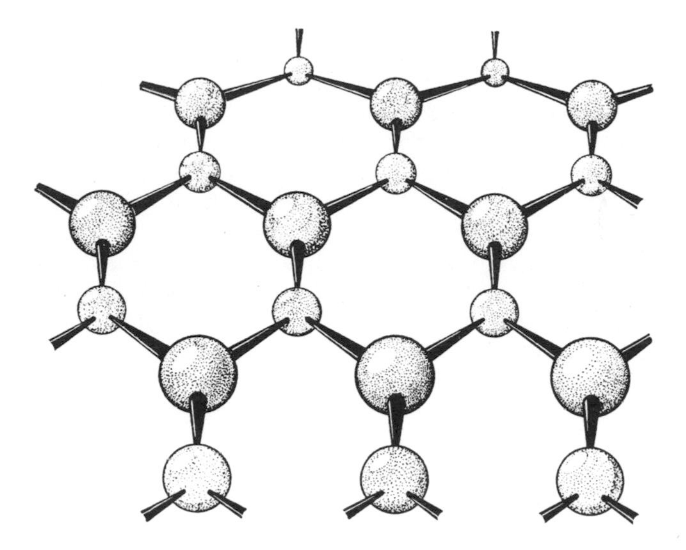

Fig. 6. Red phosphorus, puckered sheets

This model gives a plausible picture of diminished volatility and reduced solubility in covalent solvents, since the breaking up of the interlinking giant structure of atoms would require more energy than that for the simpler molecular structure of the white variety. A more marked degree of chemical inertness would also be expected from this network of covalent bonds which resists bond rupture.

The interconversion of the two allotropes of phosphorus also suggests marked structural differences. The white variety on prolonged exposure to light will undergo an allotropic change to red phosphorus, suggesting that the electron arrangement in the white allotrope is readily susceptible to change to a more stable arrangement. The change is catalysed by a trace of iodine and accelerated by heating to 523 K.

Red phosphorus on the other hand, cannot change directly while still in the solid state, to white phosphorus, though if red phosphorus is melted or vaporized and then cooled rapidly, the relatively simple white structure of P_4 tetrahedra held together by weak van der Waals' forces is formed. This suggests that the building up of the more complex red layer structure is more slowly achieved and requires rather more energetic conditions for its formation, though it is ultimately the more stable of the two arrangements.

The white–red allotropic system of phosphorus is described as *monotropic*. This means that there is no temperature (*the transition temperature*) at which both forms can coexist in equilibrium; one form, in this case red phosphorus, is more stable over the whole temperature range over which the two allotropes exist. That it has a lower energy content can be shown from calculations based on the heats of combustion of the two modifications.

2.6 Sulphur

The electron configuration is (Ne)$3s^2 3p^2 3p^1 3p^1$ with two unpaired electrons from which a covalency of two can be predicted. Just as phosphorus in the second period was different from nitrogen in not being diatomic, so sulphur differs from oxygen. Molecular weight determinations suggest S_8 as the basic molecule. Each sulphur atom pairs an electron with a neighbour on either side to form an eight-membered ring. The ring puckers to be as free from strain as possible (Fig. 7).

Sulphur shows allotropy forming monoclinic or rhombic crystals depending upon the conditions. Rhombic sulphur forms from cold solutions in covalent solvents such as carbon disulphide, and has a density of 2·07 g cm^{-3}. It can be represented as stacked S_8 rings with each ring slotted into the depressions in the ring below (Fig. 7). When rhombic crystals are heated this structure can be imagined to collapse until the S_8 rings are free to move over one another giving a straw-coloured melt at 386 K. Sulphur in this form is sometimes referred to as λ sulphur. On raising the temperature the rings break open to form linear molecules which link and cross-link, to give molecular weights corresponding to S_{20000} and resulting in an increase in viscosity and a darkening in colour. This process is complete by about 453 K and the product is known as μ sulphur. At higher temperatures, nearing the boiling point (718 K), the additional energy is sufficient to fracture the giant, intertwined chains of sulphur atoms into simpler units corresponding to S_{1000} as evidenced by a decrease in viscosity. S_8 rings are again found in the vapour phase.

In liquid sulphur a number of different forms of the element coexist in equilibrium in the same physical state, and such a system is an example of *dynamic allotropy*. The proportions of the different forms vary with temperature, and hence the properties of the melt change.

If molten sulphur is poured into cold water, the equilibrium is frozen in the chain condition favoured by the temperature of the melt, and a pliable, elastic form of sulphur, plastic sulphur, results. The plasticity is a consequence of the chains sliding over each other (Fig. 7).

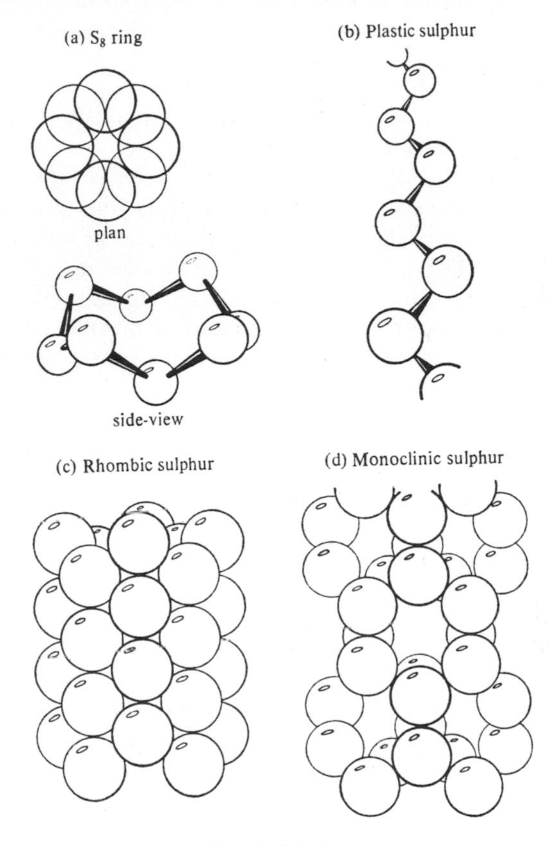

Fig. 7. Sulphur

Monoclinic crystals are formed from melts or solutions at temperatures above 369 K, the transition temperature. These crystals are less dense (density = $1{\cdot}96\ g\ cm^{-3}$) than the rhombic form and it is possible that the S_8 rings are stacked in a more open arrangement. Any form of sulphur slowly rearranges on standing to the more stable rhombic condition.

The type of allotropy shown here contrasts with that exhibited by phosphorus. The two allotropic modifications can coexist in equilibrium at the transition temperature. This type of allotropy is referred to as *enantiotropy*. Above the transition temperature the more open monoclinic structure is the stable modification, whereas below this temperature the closer packed rhombic variety becomes the stable form.

From a structural point of view it seems reasonable to suppose that the two allotropes are interconvertible. Both have as their basic structural unit the S_8 ring and the two modifications can be arrived at by packing this basic unit in different ways. No major bond breaking or bond making is required in this relatively simple rearrangement of octatomic molecules. Energy requirements are not therefore likely to be great.

Similarly, large differences in the properties of the two allotropes would not be expected. Both forms have pnysical properties typical of a molecular structure, and are soluble in a good number of covalent solvents. In terms of chemical properties, little difference is to be expected. The S_8 ring is fairly readily opened as indicated during the melting of sulphur and this is presumably a preliminary step in most reactions. It might be speculated that the monoclinic structure is more open to attack through penetration of reagents into the lattice and possibly could have slightly enhanced reactivity.

2.7 Carbon

Carbon has two allotropic forms, diamond, in which the element apparently has a valency of four, and graphite, where the valency appears to be three. The electron configuration of carbon is $(He)2s^2 2p^1 2p^1 2p^0$, suggesting a valency of two with the pairing off of the two unpaired electrons. Apparently some mode of unpairing electrons, prior to bond formation, must be envisaged to give the required number of unpaired electons. If it is supposed that one of the $2s$ electrons is promoted to the vacant $2p$ orbital, $(He)2s^1 2p^1 2p^1 2p^1$ results and this would suggest quadricovalence. However, in any compound formed, such an arrangement would give three equivalent bonds in which $2p$ orbitals take part and a non-identical bond involving the $2s$ orbital. Such a picture is at variance with the facts which suggest four identical bonds. It is therefore assumed that the $2s$ orbital and the three $2p$ orbitals merge to give four new, *hybrid* orbitals, called sp^3 orbitals. The electron configuration for carbon becomes $(He)(2sp^3)^1(2sp^3)^1(2sp^3)^1(2sp^3)^1$. The orbitals are imagined to be identical in nature and directed towards the four corners of a tetrahedron with the carbon nucleus at the centre. In diamond the four sp^3 hybrid orbitals of each carbon atom overlap with those of four neighbouring carbon atoms to give a three-dimensional network (Fig. 8). This structure suggests the strength and hardness of diamond, and the unusually high melting point comes as no surprise.

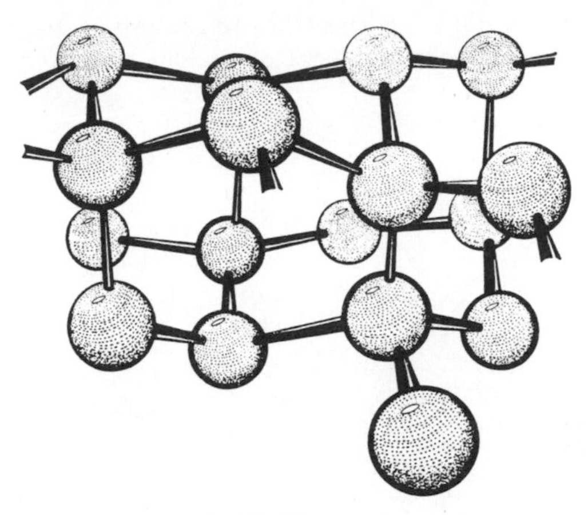

Fig. 8. Diamond

In graphite only three electrons are involved in carbon to carbon bonds. It is supposed that following promotion of a 2*s* electron to the vacant 2*p* orbital, only the 2*s* orbital and two of the 2*p* orbitals merge to give three sp^2 hybrid orbitals. In graphite then, the effective configuration of each carbon atom is $(He)(2sp^2)^1(2sp^2)^1(2sp^2)^1 2p^1$. The layer lattice structure is the consequence of $(2sp^2)$–$(2sp^2)$ overlap between carbon atoms, the angle between the bonds formed being the maximum possible angle, that is, 120 degrees. To maintain trivalence and this angle throughout the structure, a planar giant molecule composed of six membered rings is visualized (Fig. 9). If three of the four valency electrons of carbon are used in this structure, what happens to the remaining electron? It would appear to be in a *p* orbital at right angles to the plane of the molecule and centred on each carbon atom. These *p* orbitals are imagined to overlap sideways from carbon atom to carbon atom, above and below the plane of the molecule, delocalizing the electrons over the whole lattice. The electrical conductivity of graphite is explained, since the delocalized electrons can, under the influence of an electric potential, flow through the structure. The hexagonal planar networks are kept apart by the delocalized orbitals, and it is natural that they can slide over one another, helping to explain the lubricant properties of graphite.

From this structure for graphite it can be predicted that many of its physical properties would be dependent upon the direction in which they are measured. For example, properties such as tensile strength, thermal conductivity, and electrical resistivity would be expected to change if measured first parallel to, and then at right angles to, the crystal planes. Such a substance is said to be *anisotropic* in contrast to *isotropic* materials when the properties are independent of the direction in which they are measured.

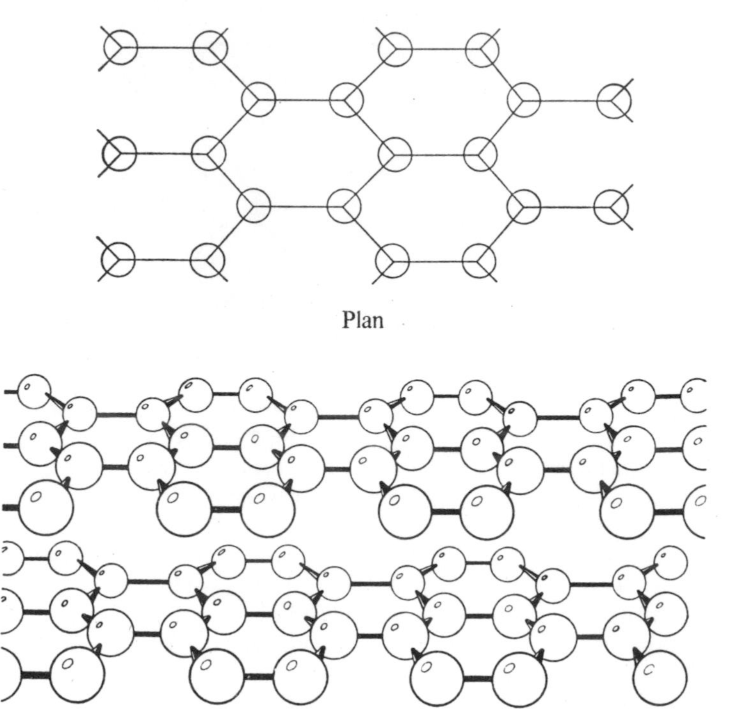

Fig. 9. Graphite

Because of its potentially anisotropic nature, research has been carried out in recent years to produce large single pieces of graphite. These investigations have led to the development of *pyrographite*, in which the tiny graphite crystals lie in the same direction. Pyrographite is highly anisotropic and has found use in space research in rocket nozzles and re-entry capsules. Its thermal conductivity is much greater along the crystal planes than at right angles to them. A layer of pyrographite can allow heat to be conducted away in one direction from high temperature areas, while its good insulating properties at right angles to this direction provide protection during re-entry.

The graphite–diamond allotropic system is another example of monotropy. Graphite is rather surprisingly the stable modification by 1·9 kJ mol^{-1} of C at room temperature and pressure, though the conversion of diamond to graphite occurs infinitely slowly under normal conditions. The difficulty of interconverting these two allotropes can be expected from a structural point of view. The breaking down and reforming of giant interlinked structures of covalent bonds are required, not as with sulphur a rearrangement of simple molecular units.

Some differences in chemical behaviour between the two allotropes might be expected. Neither of them show a high degree of chemical reactivity, and this again is associated with their giant structures. Graphite reacts more readily, and this can be explained by assuming that the layers of carbon atoms are penetrated by molecules of the reactant. It will for instance take fire in oxygen or fluorine at a lower temperature than diamond. Compounds of graphite with both metals and non-metals, in which the reacting atoms or molecules go in between the layers of carbon atoms are also known; a mode of reaction not open to diamond.

3 More Covalent Compounds

A covalent bond is, as we have seen, a shared pair of electrons, and when the sharing atoms are non-identical the nature of the bond will be modified by the difference in the power of attraction for electrons of the two atoms.

This power of an atom or molecule to attract electrons to itself is termed *electronegativity*. An arbitrary scale of electronegativity values has been devised by Pauling from calculations using bond energies (see Table 6). Other workers have devised other methods of calculation. One of the simplest, for example, employs the formula:

$$\text{electronegativity} = \frac{\text{effective nuclear charge}}{\text{covalent atomic radius}}$$

It is important to distinguish between electronegativity and *electron affinity*. The latter is a measure of the energy change in the reaction

$$X_{(g)} + e \rightarrow X^{-}_{(g)}$$

in which X represents an unbound atom in the gas phase and e an electron. In other words, electron affinity is the ionization energy for a negative ion.

The higher the electronegativity value, the greater the attraction of an element for the shared electrons in a bond. The difference in electronegativities for two sharing elements is a measure of bond distortion.

Table 6. Electronegativity values for some elements (PAULING, L. *The Nature of the Chemical Bond*, 3rd edn. O.U.P., 1960, p. 90)

H 2·1			
C 2·5	N 3·0	O 3·5	F 4·0
Si 1·8	P 2·1	S 2·5	Cl 3·0
Ge 1·8	As 2·0	Se 2·4	Br 2·8
			I 2·5

If two elements A and B, where B has the higher electronegativity, share a pair of electrons, one from each atom, the end of the molecule in the neighbourhood of B will be negative with respect to end A. Such an arrangement is termed a *dipole*, and the extent of unequal charge distribution is given by the *dipole moment* (p). Dipole moment is defined as the product of the excess charge on one atom and the internuclear separation

$$p = q \times d$$

where q is the excess charge in coulombs and d the equilibrium distance between the two nuclei in metres. Since the molecule as a whole is electrically neutral, the excess charge on the one atom will be balanced by an equal charge of opposite sign on the other. Dipole moments can be estimated from the tendency of the molecules of a compound to line up in an electric field. The greater this tendency, the greater the dipole and the more extreme the inequality of charge distribution.

Table 7. Dipole moments for some compounds

Compound	p/C m
HF	$6{\cdot}42 \times 10^{-30}$
HCl	$3{\cdot}46 \times 10^{-30}$
HBr	$2{\cdot}66 \times 10^{-30}$
HI	$1{\cdot}28 \times 10^{-30}$
H_2O	$7{\cdot}61 \times 10^{-30}$
H_2S	$3{\cdot}13 \times 10^{-30}$
CO_2	0·00
NH_3	$4{\cdot}84 \times 10^{-30}$
NF_3	$0{\cdot}67 \times 10^{-30}$
PH_3	$1{\cdot}85 \times 10^{-30}$

3.1 Hydrogen halides

The hydrogen halides are asymmetrical covalent molecules. In each of them, the covalent bond is the result of overlap between the 1s orbital of the hydrogen atom with a p orbital of the halogen (Fig. 10). This gives a structure in which all orbitals are paired and the sharing atoms attain, in effect, noble gas configurations.

Unlike the substances discussed in Chapter 2, the sharing atoms have unequal electronegativities and dipoles are formed. The electronegativity difference decreases in the order F, Cl, Br, I, and the dipole moments reflect this fact. The dipole in hydrogen chloride is caused by the greater attraction of the chlorine atom for the shared electrons. In hydrogen bromide, the increased size of the bromine atom means that the bromine nucleus is further from the area of overlap with hydrogen, its attractive force is less and the dipole moment is smaller. This situation is exaggerated in the case of hydrogen iodide. In hydrogen fluoride, on the other hand, the nucleus of the small fluorine atom is nearer to the electrons of the bond, distortion is more extreme and the dipole moment is larger.

Consideration of the chemistry of the hydrogen halides supports the view that hydrogen chloride, hydrogen bromide and hydrogen iodide reflect in a

Hydrogen fluoride seems to be peculiar. It is formed very readily under low energy conditions and the reaction is markedly exothermic.

The ease of initiation of the reaction is associated with the abnormally low F—F bond energy, which is comparable with that of the I—I bond in I_2. The weakness of the F—F bond is rather difficult to account for. Several possible explanations have been put forward. One of these suggests that since the F—F bond is very short (the fluorine atoms are small), repulsive forces are set up between the non-bonding electrons on the two atoms, resulting in a weakening of the bond. Another suggestion is that some form of multiple bonding occurs in the other halogens which is not available to fluorine since it cannot expand its octet.

The high degree of exothermicity is mainly associated with the formation of the very strong H—F bond. The hydrogen fluoride molecule shows abnormal thermal stability and the intermolecular forces of attraction are apparently exceptionally high. There is evidence that liquid hydrogen fluoride consists of a zig-zag chain polymer (Fig. 11) and that the species H_2F_2,H_4F_4, and even H_6F_6, exist in the vapour at moderate temperatures. It is not until 373 K that the vapour is monomolecular. Liquid hydrogen fluoride is, like water, an ionising solvent. All these effects are due to the high degree of distortion in the molecule.

Extremes of dipole–dipole attraction where some degree of structure results is sometimes referred to as *hydrogen bonding*.

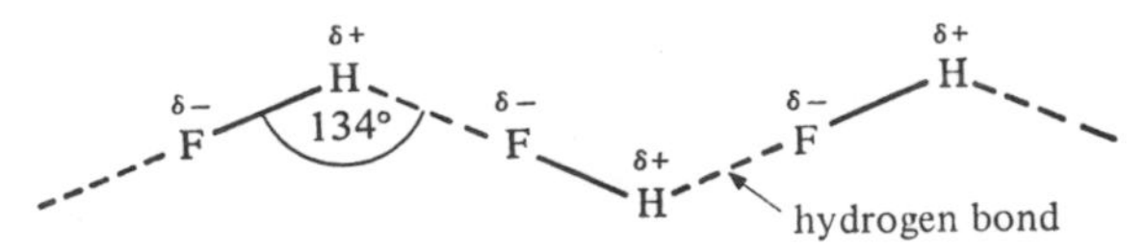

Fig. 11. Chain polymer in liquid hydrogen fluoride

3.2 Water

The dipole moment of water is $7{\cdot}61 \times 10^{-30}$ C m, of the same order as that of hydrogen fluoride, so it is perhaps less surprising that the properties of water are exceptional. Figure 12 shows that the oxygen atom shares and pairs the electron in each of its half-filled 2*p* orbitals with the electron in the 1*s* orbital of the two hydrogen atoms. Since the 2*p* orbitals are at right angles, this model predicts a bond angle of 90 degrees for the H—O—H molecule. In each O—H bond the greater electronegativity of the oxygen distorts the bond, the oxygen end being negative with respect to the hydrogen. According to this model, there will be two dipoles at right angles and these give to the molecule as a whole, a resultant dipole.

Although this simple interpretation does explain the existence of a dipole on the water molecule, the predicted bond angle of 90 degrees differs from the experimentally observed 104 degrees 30 minutes. It is sometimes suggested that the repulsion between the electrons in the two bonding orbitals accounts for the difference, and a scale drawing of a water molecule gives some substance to this idea.

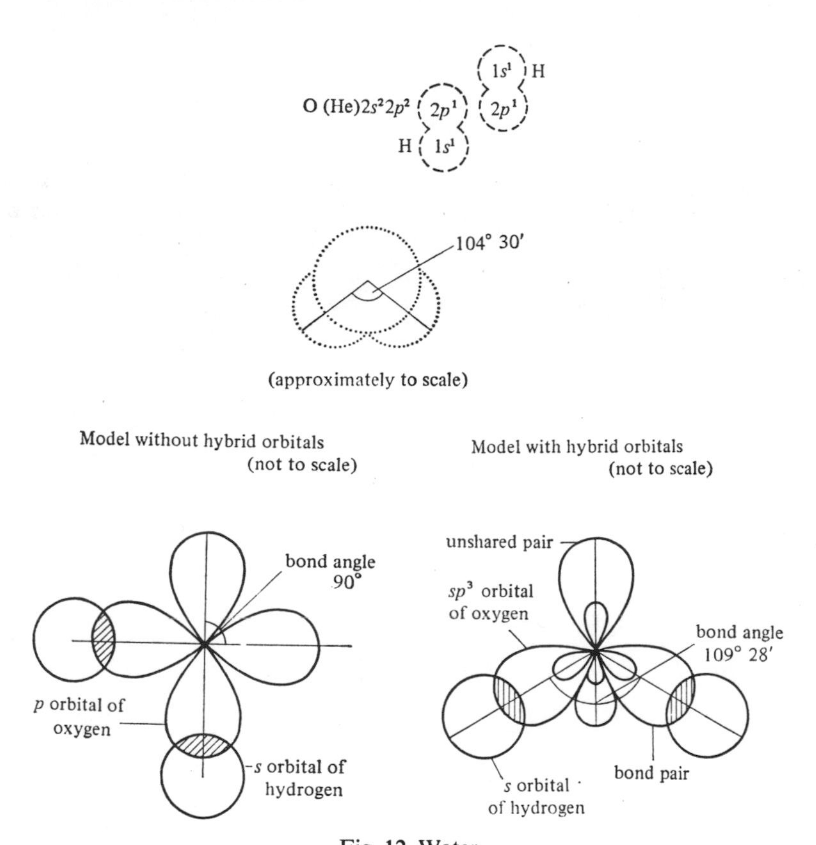

Fig. 12. Water

An alternative explanation is to suppose that the electrons in the 2*s* and two 2*p* orbitals of the oxygen atom occupy four hybrid orbitals, in the manner assumed for carbon. These four hybrid orbitals are directed towards the corners of a tetrahedron and if two of them overlap with the 1*s* orbital on two hydrogen atoms this would give a bond angle of 109 degrees 28 minutes. It has also been suggested that unshared pairs of electrons repel one another with a greater force than they do bond pairs, and this force is, in turn, greater than the repulsion between bond pairs. This notion gives a reason for a reduction in the bond angle of water from the ideal tetrahedral angle of 109 degrees 28 minutes.

The structure of ice (Fig. 13) seems to give support to the tetrahedral idea, since each oxygen atom appears to be surrounded by four hydrogen atoms in a three-dimensional network. Two of the hydrogen atoms are covalently bonded to the central oxygen atom, while the other two are held in place by dipole–dipole attractions like the 'hydrogen bond' noted in hydrogen fluoride.

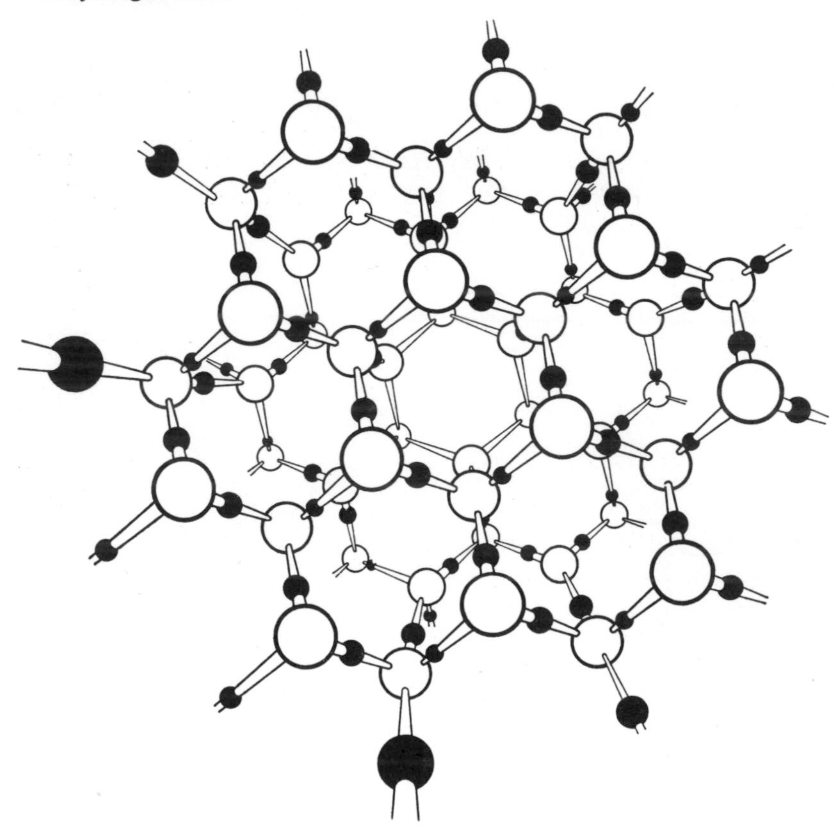

Fig. 13. Ice

For ice to melt, large numbers of attractions between dipoles have to be overcome and the heat of fusion (6·03 kJ mol^{-1}) is accordingly high. On melting, the open network structure collapses, the water molecules crowd closer together and the density increases. Collapsing predominates until a temperature of 277 K is reached, after which the separation of water molecules plays the major part and the volume increases, accounting for the maximum density of water at 277 K. Increasing the temperature further, increases the average kinetic energy of the water molecules and hence, the average distance between the molecules. More and more molecules acquire sufficient energy to overcome the forces between molecules within the liquid, escape through the surface of the water, and the rate of evaporation increases. The aggregations of molecules, resulting from dipole–dipole attractions, become simpler until the stage is reached where molecular separation is sufficient to form vapour within the body of the liquid, and the water boils.

Water is a good ionizing solvent and as we shall see later, the dipole moment and unshared electron pairs give an explanation of the fact. It might be remarked in passing that a linear model for water would have a zero resultant dipole moment and could not account for its properties.

Gramme for gramme, water has a higher heat absorption capacity than most substances, as evidenced by the relatively high specific heat (4·18 $kJ\ kg^{-1}\ K^{-1}$). This is in harmony with the structure, since heat energy is absorbed in overcoming dipole–dipole attractions between molecules, and in keeping them apart, rather than in increasing the average kinetic energy and hence the temperature.

Water is a stable compound reflecting the strength of the O—H bond, which is remarkably resistant to thermal rupture. Even at 2300 K only 2 per cent of water molecules are decomposed.

In terms of its chemical properties the presence of the two lone pairs on the molecule is important. These unshared pairs can be donated to a suitable acceptor molecule or ion to give a temporary or permanent bond. For instance, in the hydrolysis of covalent halides a preliminary step generally involves the formation of a bond between the water molecule and the atom attached to the halogen through the use of a lone pair on the oxygen. The halides of the first-row non-metallic elements such as CCl_4 and NF_3 are relatively resistant to hydrolysis since the halogenated atoms have no orbitals of suitable energy available and cannot therefore expand their octets to form a temporary bond with a water molecule.

3.3 Hydrogen sulphide

The bond formation in hydrogen sulphide corresponds to that of water, the two unshared $3p$ electrons of sulphur pairing off with the $1s$ electrons from two hydrogen atoms (Fig. 14). Recalling that the three p orbitals are mutually at right angles a bond angle of 90 degrees is to be expected. This compares well with an observed angle of 92 degrees, and there is no need to postulate hybrid orbitals.

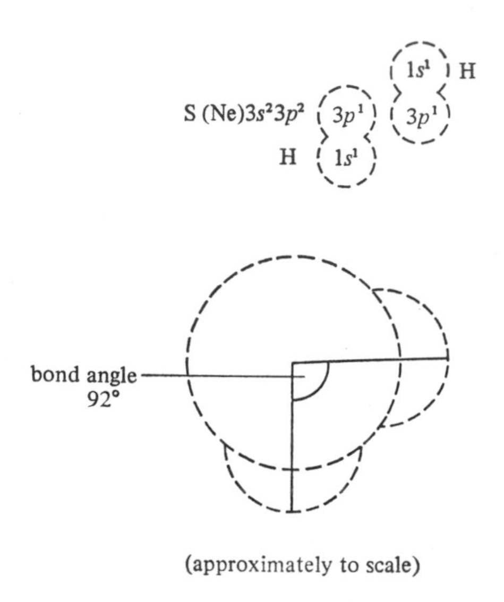

(approximately to scale)

Fig. 14. Hydrogen sulphide

Compared with water, the orbital overlap will be less since the larger sulphur atom exerts a smaller electron attracting effect. This view is supported by an electronegativity difference between hydrogen and sulphur of 0·4, compared with 1·4 between hydrogen and oxygen. In consequence, the dipole moment of hydrogen sulphide, $3{\cdot}13 \times 10^{-30}$ C m, is much lower than that of water, $7{\cdot}61 \times 10^{-30}$ C m, and the boiling point, 211 K is correspondingly lower despite the increased molecular size. The S—H bond is weaker and hydrogen sulphide is more reactive than water, as illustrated by the ease with which sulphur is deposited in an oxidative environment.

Figure 15 shows that the boiling points of the group VI hydrides increase from sulphur to tellurium, despite the fact that the electronegativity differences between the elements and hydrogen, and hence the dipole moment, decrease. This can be explained by the increase in the magnitude of the van der Waals' forces, as the electron population of the molecules increases. The relative boiling points of the group VI hydrides mirror those of group VII, the first member of each group showing abnormality as a result of hydrogen bonding, the others reflecting the fact that dipole–dipole attractions become less significant than van der Waals' forces.

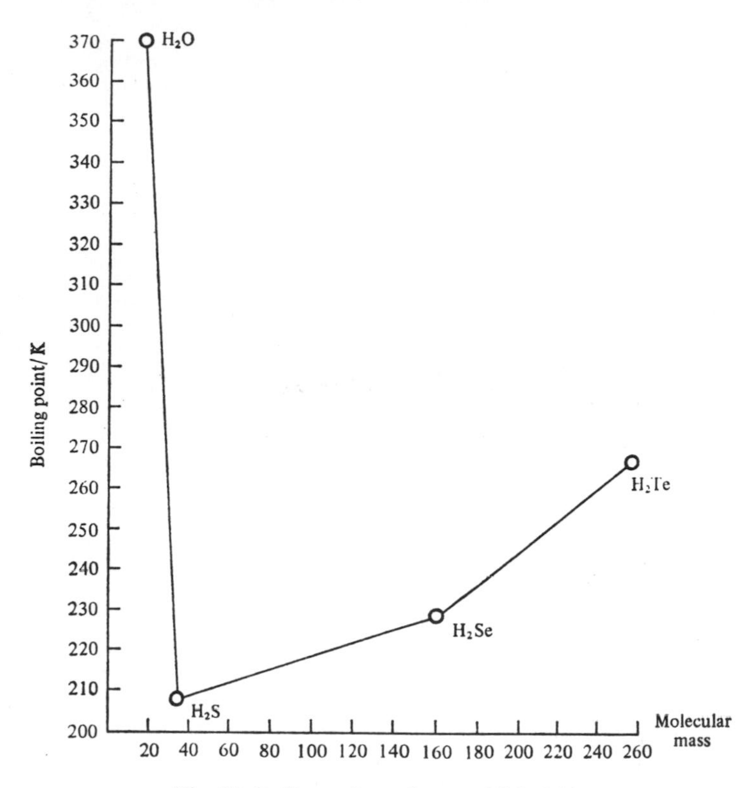

Fig. 15. Boiling points of group VI hydrides

3.4 Hydrogen peroxide

From the electron configurations of hydrogen and oxygen, the necessary sharing of unpaired electrons leads to the arrangement shown in Fig. 16. The three-dimensional structure is the consequence of repulsions between the bond pairs and the unshared pairs of electrons.

The molecule is polar, dipole moment 7·16 C m, and pure hydrogen peroxide is a syrupy liquid which freezes at 271 K and boils at 431 K. Dipole–dipole attractions with water molecules lead to complete miscibility between the two liquids. The dominant feature of the chemistry of hydrogen peroxide is its tendency to give the more stable water, either liberating gaseous oxygen or oxidizing other substances. This behaviour is associated with the weakness of the O—O bond.

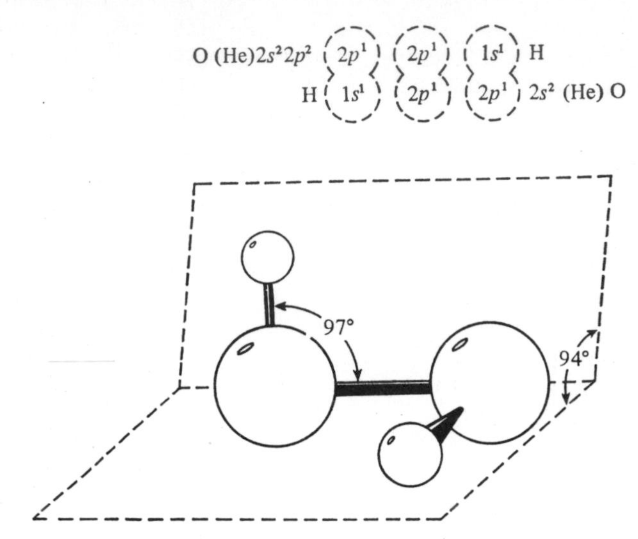

Fig. 16. Hydrogen peroxide

3.5 Ammonia

In ammonia, the nitrogen atom pairs off its three unpaired electrons with the electrons in each of three hydrogen atoms. Nitrogen has a configuration $(He)2s^22p^12p^12p^1$, with the three p orbitals mutually at right angles. The

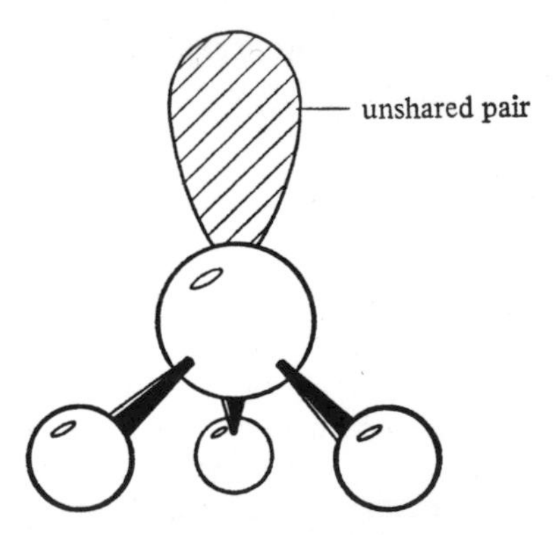

Fig. 17. Ammonia

ammonia molecule can be expected to be tetrahedral in shape with a nitrogen atom and three hydrogen atoms at the four corners and a bond angle of 90 degrees. However, the measured bond angle is 106 degrees 45 minutes and we could attempt to account for the difference by assuming repulsions between the unshared pair of electrons on the nitrogen atom and the electron pairs in the bonds, or by postulating hybrid orbitals, or by a combination of both these ideas.

In a model involving hybrid orbitals the projecting orbital of the unshared pair of electrons on the nitrogen atom will, as we shall see, give a plausible picture of the electron donor properties of ammonia. An unhybridized model would not account for this as the unshared electrons would be supposed to be in a spherically symmetrical 2*s* orbital, which would not project.

The electronegativity difference between nitrogen and hydrogen is 0·9 and the three dipoles give a resultant of $4{\cdot}84 \times 10^{-30}$ C m. Liquid ammonia is like water, a good ionizing solvent. The boiling points of the group V hydrides show the same trends as were noted for the hydrides of groups VI and VII, but ammonia is not, relatively, as abnormal as are water and hydrogen fluoride.

3.6 Nitrogen trifluoride

Nitrogen trifluoride gives interesting support for the idea of hybrid orbitals. In the molecule overlap between the central nitrogen atom and the three fluorine atoms results in electrons being paired throughout. The electronegativity difference between nitrogen and fluorine is 1·0 suggesting a high

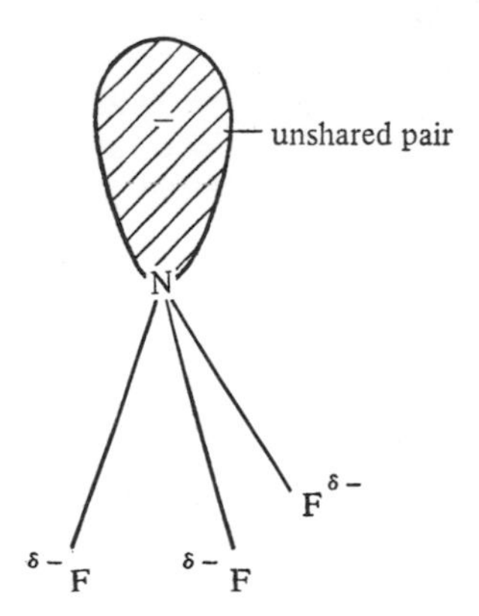

Fig. 18. Nitrogen trifluoride

resultant dipole moment for the molecule. However, this value is $0{\cdot}67 \times 10^{-30}$ C m and this is understandable if it is supposed that the orbital, occupied by an unshared electron pair, opposes the resultant polarity of the N—F bonds (Fig. 18).

The resistance of nitrogen trifluoride to hydrolysis has already been mentioned. Neither nitrogen nor fluorine are able to expand their octets, and therefore cannot accept a lone pair from a water molecule to give the initial step in the hydrolysis.

Nitrogen trichloride is slowly hydrolyzed by water, and this presumably involves the expansion of the chlorine octet.

3.7 Phosphine

Phosphorus, with a configuration of (Ne)$3s^2 3p^1 3p^1 3p^1$, can pair off electrons with three hydrogen atoms. The bond angle of 93 degrees 50 minutes is reasonably near to the anticipated 90 degrees for three *p* orbitals, the difference could be the result of electron repulsion between the electron clouds of the three hydrogen atoms, but hybrid orbitals may make a contribution. The electronegativity difference between phosphorus and hydrogen is nil so the fact that the molecule has a dipole moment of $1{\cdot}85 \times 10^{-30}$ C m suggests that the unshared pair of electrons on the phosphorus atom must project to some extent.

However, there is little tendency to produce association through hydrogen bonding, so the melting point, boiling point and other physical properties of phosphine are much more compatible with its molecular weight than are those of ammonia.

From a chemical point of view phosphine is much less stable than ammonia. It is more reactive in most circumstances, and will burn readily in air. The P—H bond is weaker than the N—H bond and this contributes to the enhanced reactivity. One other difference is that the phosphorus atom in phosphine is a less effective donor of electrons than the nitrogen atom in ammonia. It is for this reason more feebly basic than ammonia (see 5.4).

3.8 Phosphorus trichloride

Phosphorus trichloride is similar to phosphine in structure with three-fold *p–p* overlap. The electronegativity difference between phosphorus and chlorine is $3{\cdot}02 \times 10^{-30}$ C m and the resulting dipole moment is suggestive of the reactivity of phosphorus trichloride with water, another polar molecule. The hydrolysis of halides of this type has already been mentioned.

4 Ionic Compounds

Looking at the electron configurations of the alkali metals alongside those of the halogens suggests another method of electron rearrangement, which would give a noble gas configuration (Table 9).

Table 9

Alkali metals		*Halogens*	
Li	(He)$2s^1$	F	(He)$2s^2 2p^2 2p^2 2p^1$
Na	(Ne)$3s^1$	Cl	(Ne)$3s^2 3p^2 3p^2 3p^1$
K	(Ar)$4s^1$	Br	(Ar)$3d^{10} 4s^2 4p^2 4p^2 4p^1$
Rb	(Kr)$5s^1$	I	(Kr)$4d^{10} 5s^2 5p^2 5p^2 5p^1$
Cs	(Xe)$6s^1$		
M	(noble gas)s^1	X	(inner electrons)$s^2 p^2 p^2 p^1$

If an alkali metal atom loses an electron the resulting particle will have a noble gas configuration; if a halogen atom accepts an electron it too will form a more stable arrangement. The loss or gain of an electron produces an *ion* of unit positive or unit negative charge.

$$\text{M(noble gas)}s^1 - 1e = \text{M(noble gas)}^+$$

$$\text{X(inner electrons)}s^2p^2p^2p^1 + 1e = \text{X(noble gas)}^-$$

An electron is transferred from the *s* orbital of an alkali metal atom to the *p* orbital of a halogen atom.

The elements of the alkaline earth group have two electrons in the *s* orbital and by losing them to an electron acceptor, such as a halogen, form bi-positive ions.

$$\text{X(inner electrons)}s^2p^2p^2p^1 + 1e = \text{X(noble gas)}^-$$

$$\text{M(noble gas)}s^2 - 2e = \text{M(noble gas)}^{2+}$$

$$\text{X(inner electrons)}s^2p^2p^2p^1 + 1e = \text{X(noble gas)}^-$$

The atoms of oxygen and sulphur are two electrons short of neon and argon configurations respectively, and by gaining two electrons attain more stable arrangements. Similarly a hydrogen atom by accepting one electron becomes H^-, with a helium configuration.

It will be recalled that in discussing the electron arrangements in Chapter 2, the reactivity of molecules, already having a noble gas configuration, was explained in terms of rearrangements of electrons to give structures of greater stability. It can now be seen that the formation of ions may be such a rearrangement. The formation of ions can be thought of as the extreme case of unequal sharing where the more electronegative element has gained complete control of both electrons.

4.1 Factors affecting the formation of ions

The energy required to remove an electron from a neutral atom decreases with the increase in distance between the nucleus and the electron. In consequence the ionization energy decreases down a group of the periodic table as the atomic size increases.

Once one electron has been removed from an atom, the nucleus attracts the remaining electrons with a greater attractive force than before and the second ionization energy is greater than the first.

Size is not the only factor influencing ionization energies. The electrons between the nucleus and those to be lost act as a shield and reduce the force of attraction. The greater the electron population in this inner zone, the greater this so called, *screening effect.* This additional factor reinforces the influence of size and helps to lower the ionization energies of atoms with greater atomic numbers.

If noble gas configurations are to be formed by electron gain, smaller atoms act as acceptors more readily than larger ones, since the nucleus is nearer to the shell which accepts the incoming electron. Once one electron has been accepted it becomes more difficult to introduce a second against the repulsion between particles of like charge. The screening effect results in atoms of higher atomic number being poorer acceptors and again reinforces the size factor.

Fajans' rules summarize the conditions favouring the formation of ions. An ion forms more readily when:

a. the electronic structure of the ion is stable

b. the charge on the ion is small

c. either the atom which forms the ion is small if negative ions are formed or the atom which forms the ion is large if positive ions are formed.

4.2 Born–Haber cycle

To understand the formation of ionic compounds from the parent elements, it is necessary to consider a number of energy terms, and these are summarized in a *Born–Haber cycle.*

For the formation of a stable ionic compound the heat of formation Q, must be negative. That is, taken over all there must be a liberation of energy with respect to the starting materials. The greater the negative value of Q, the more stable the crystal lattice produced.

Table 10. Born–Haber cycle

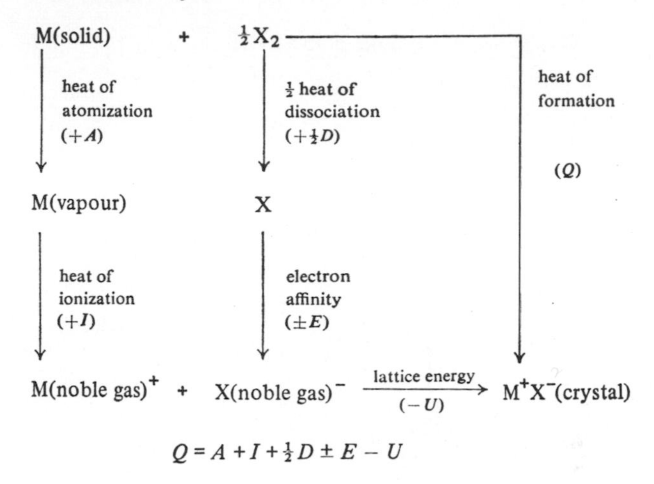

$$Q = A + I + \tfrac{1}{2}D \pm E - U$$

4.3 The terms *A* and *I* (Table 11)

In a metallic element the atoms are closely packed in the most stable arrangement, and may be thought of as being held together by bonds between the atoms. Before electrons can be removed to form ions, the forces of attraction between atoms in the solid metallic element will have to be overcome, a process requiring the removal of atoms to the gas phase so that they are independent of one another. The energy required is the *heat of atomization* $(+A)$.

Further energy is needed to remove electrons from the atoms, to form positive ions, and the ionization energy $(+I)$ is a measure of this.

4.4 The terms *D* and *E* (Table 12)

Non-metallic elements consist of covalently bonded atoms, and energy will be needed to break these bonds before the resulting atoms can accept electrons to form ions. The energy needed to dissociate the molecules of a non-metallic element into atoms is the *heat of dissociation* $(+D)$, and the energy change resulting from electron acceptance, in the formation of negative ions, is the electron affinity $(\pm E)$. In practice, E is often difficult to measure; a calculated value for the lattice energy may be used in the Born–Haber cycle to evaluate E or some other estimation procedure used.

Reference to Table 12 shows that the formation of univalent negative ions is accompanied by heat evolution, but for bivalent negative ions

Table 11. Ionization energies and heats of atomization

Element	*1st ionization energy* ($kJ\ mol^{-1}$)	*2nd ionization energy* ($kJ\ mol^{-1}$)	*3rd ionization energy* ($kJ\ mol^{-1}$)	(I) ($kJ\ mol^{-1}$)	(A)
Li	+ 519			+ 519	+ 161
Na	+ 494			+ 494	+ 109
K	+ 418			+ 418	+ 90
Be	+ 900	+ 1757		+ 2657	+ 321
Mg	+ 736	+ 1448		+ 2184	+ 150
Ca	+ 590	+ 1146		+ 1736	+ 193
Al	+ 577	+ 1816	+ 2753	+ 5146	+ 314

energy is required. This is because if an atom accepts one electron to form a negative ion of unit charge, a second electron can only be introduced against the repulsion between the ion and the incoming electron. For example, one mole of oxygen evolves 138 kJ in forming O^- ions, but 791 kJ are required to introduce the second electron and form O^{2-} ions. Overall, 653 kJ mol^{-1} of O are required in forming bi-negative oxygen ions from the atoms.

Table 12. Electron affinities and heats of dissociation

Element	*Electron affinity* (E) (kJ mol^{-1})	*Heat of dissociation* (D) (kJ mol^{-1})
F	− 349	+ 77
Cl	− 365	+ 121
Br	− 343	+ 112
I	− 317	+ 107
H	− 74	+ 218
O	+ 653	+ 248
S	+ 335	+ 219

4.5 The terms U and Q (Table 13)

When the ions in the gaseous state come together to form a crystal lattice, the energy evolved is referred to as the *lattice energy* ($-U$). Its direct determination is not possible and a Born–Haber cycle can be used to obtain it. The sum of the energy terms A, I, $\frac{1}{2}D$, and E for a number of ionic compounds has a positive value, that is the ionic compounds would not be stable, if they formed at all, but for the contribution of the lattice energy.

Table 13. Lattice energies and heats of reaction

Compound	$A = I + \frac{1}{2}D \pm E$ (kJ mol^{-1})	*Lattice energy* (U) (kJ mol^{-1})	*Calculated heat of formation* (Q) (kJ mol^{-1})
LiF	+ 368	− 1021	− 652
NaF	+ 291	− 900	− 609
KF	+ 197	− 795	− 598
LiCl	+ 375	− 845	− 470
NaCl	+ 298	− 770	− 472
KCl	+ 204	− 699	− 495
NaBr	+ 315	− 732	− 417
KBr	+ 221	− 669	− 448
$MgBr_2$	+ 2105	− 2406	− 301

The *heat of reaction*, Q, is a measure of the stability of an ionic compound. The greater the negative value of Q, the more likely its formation and, once formed, the greater its stability relative to the starting materials. In general terms the value of Q is governed by the relative magnitudes of U and I. These two terms make the largest contributions to a cycle and are always opposite in sign. If less energy is required to ionize the metal atoms than is returned by the formation of the crystal lattice, then the compound can be expected to form and vice versa.

It should be noted that although the foregoing discussion is useful in predicting the likelihood of formation of ionic compounds, and is capable of extension to predict when compounds will not form, e.g. AlF, $CaCl_3$, $NaCl_2$ and so on, it is both a simplification and an approximation. The various energies are known only approximately and the inherent assumption that the compounds are 100 per cent ionic is rarely, if ever, completely valid. This limitation is important, and a Born–Haber cycle can be used to give an indication as to how far our ideas of the bonding in a particular compound are valid. If we assume a material consists of an array of discrete spherical ions in contact with one another, it is possible to calculate the energy given out in forming this lattice when widely separated charged particles are brought together. In other words, a calculated value for the lattice energy can be obtained for comparison with a value obtained by using a Born–Haber cycle.

In the case of the alkali–metal halides, agreement within 1 per cent is found between the two values. Our ionic model for these compounds is therefore supported. When a similar procedure is adopted for the silver halides however, discrepancies of about 15 per cent are found and the bonding situation in these compounds is unlikely to be as markedly ionic.

4.6 Origin of lattice energy

The force which holds the ions together in a lattice is the attraction between charged particles of opposite sign. Assuming the ions to be spherical and of definite size, the factors which determine the magnitude of this force are the size of and charge on the ions. In the crystal lattice the ions are arranged to accommodate as many ions in contact with a central ion as possible, to give the most closely packed and stable structure. The number of ions surrounding a given ion is termed the *coordination number* and depends upon the ratio of cation to anion size (Table 14). Anions are generally larger than cations, but the nearer to being equal the ion sizes become, the greater the coordination number. The value of the coordination number determines the arrangement of ions around one another and hence, the crystal type.

Table 14. Radius ratio and coordination number

Cation size/anion size	*Coordination number*	*Spatial arrangement about central ion*
0·155 to 0·225	3	Planar triangular
0·225 to 0·414	4	Tetrahedral
0·414 to 0·732	6	Octahedral
Above 0·732	8	Cubic

4.7 The alkali metal halides

The radius ratio of most alkali metal halides lies between 0·44 for lithium fluoride, and 0·69 for rubidium iodide, leading to a coordination number of 6 and an octahedral arrangement of ions, as illustrated by the structure of

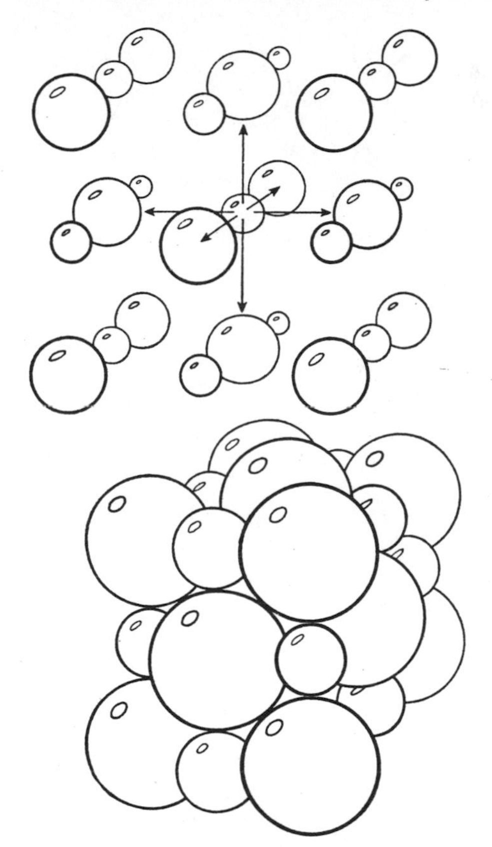

Fig. 19. Sodium chloride

sodium chloride. The radius of a sodium ion is 0·095 nm and that of a chloride ion 0·018 nm. The ratio of the radius of a sodium ion to that of a chloride ion is thus about 0·5. In the model for sodium chloride (Fig. 19), a cubic face-centred arrangement of chloride ions and a similar arrangement of sodium ions are interlocked.

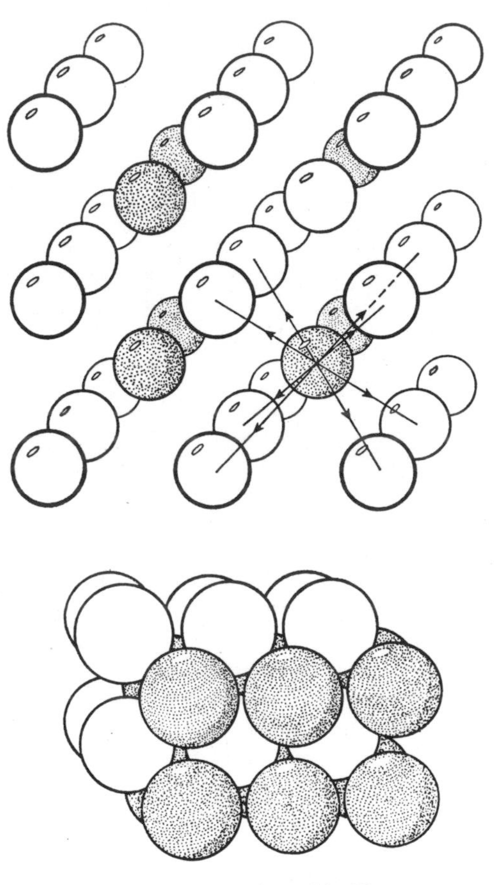

Fig. 20. Caesium chloride

If the radius ratio exceeds 0·732 each ion can be expected to be surrounded by eight oppositely charged ions and this is illustrated by caesium chloride (Fig. 20). In this model a cubic arrangement of chloride ions interlock with a cubic arrangement of caesium ions.

According to their radius ratios, lithium chloride (0·33), bromide (0·31), and iodide (0·28) should have a tetrahedral arrangement of ions. Instead, they are found to be similar to sodium chloride. It has been suggested that deviations from ionic bonding and inaccuracies in the calculation of ionic radii are possible reasons.

4.8 Melting and electrolysis of melts

As indicated by the lattice energies, the energy required to overcome the forces of attraction between ions in solid ionic compounds is considerable. Melting points of ionic compounds are high, reflecting the resistance to crystal collapse.

The ions in a molten ionic compound are free to move and electrical conductance becomes possible. At the cathode the metal ions gain electrons and the free metal is formed; at the anode negative ions lose electrons, liberating free atoms which combine to give molecules of the non-metal. The drift of ions under the influence of an electric potential, through the melt, serves to complete the circuit, carry the current, and electrolysis takes place.

4.9 Solubility of ionic compounds in water

When an ionic solid dissolves in water, the forces of attraction between the ions are overcome, the solid collapses and the individual ions, surrounded by a sheath of water molecules, are free to move. In a way the process of dissolving can be likened to melting, but the energy needed to separate the ions arises from the force of attraction between the water molecules and the ions.

At the surface of a crystal, projecting ions are pulled upon by water dipoles, dislodging them from the lattice. Other ions are exposed to the attractions of the water molecules and the process continues. An ion once removed from the lattice, is surrounded by layers of water molecules; it is said to be *solvated.* As a result of ion-dipole attractions the energy required to disintegrate the crystal is available. This view of the mechanism of dissolution of ionic compounds suggests that all ionic compounds should be water soluble and this is not the case.

Comparison of heats of hydration with lattice energies shows that there is a very small difference between two large energy terms, and it is this which determines whether an ionic compound is soluble or not, and the way in which the solubility varies with temperature. It the heat liberated on solvating the ions exceeds the lattice energy the compound can be expected to be soluble, but if the converse is true it is likely to be much less soluble. It must be emphasized that this treatment is greatly oversimplified. A more elaborate discussion would take into account changes in entropy on solution, an energy factor which though relatively small becomes significant in these circumstances, and would allow for the possibility of a degree of covalent bonding within the lattice.

Water is not the only polar solvent and ionic compounds are soluble in liquid ammonia, nitromethane, liquid sulphur dioxide, and other substances with high dipolar moments. Insolubility in non-polar solvents is to be expected since no force operates to break down the lattice, nor to solvate the ions.

4.10 Rates of Ionic Reactions

The structural differences between ionic and covalent substances outlined so far explain the wide divergence of physical properties between these two extreme types of compound. One further point of difference is in the rate at which chemical reactions involving ionic materials take place as compared with those in which covalent substances take part.

A large number of ionic reactions which are dealt with in the laboratory are carried out in solution. These, for example the precipitation reactions employed in qualitative analysis, involve the aggregation of ions into a stable crystal lattice, and in general take place in very short time intervals, and many appear to be instantaneous. It is evident that such reactions require relatively easily achieved electron rearrangements, and occur via a number of steps not involving any major energy barriers.

In contrast to this, many reactions between covalent substances take place more slowly. The common ones encountered in the laboratory are those involving organic reactants. The reactants often need to be retained in the correct environment for a long time to achieve the desired reaction. The breaking and remaking of covalent bonds generally requires more complex and often more energy consuming reaction routes to attain the new electron arrangements than the typical ionic reaction.

5 Coordination Compounds

Aluminium chloride vaporizes at 463 K, is a non-conductor when molten, and fumes in moist air, hydrolyzing to aluminium hydroxide and hydrogen chloride. These properties suggest a covalent rather than an ionic structure, despite the fact that ions of a noble gas configuration would result from electron transfer:

$$Al - 3e \rightarrow Al^{3+} \quad 3Cl + 3e \rightarrow 3Cl^{-}$$

The resulting ions could presumably form an ionic lattice with liberation of energy. That this does not happen, may be a consequence of the excessive energy 5146 kJ mol^{-1}. needed to form aluminium ions from aluminium atoms.

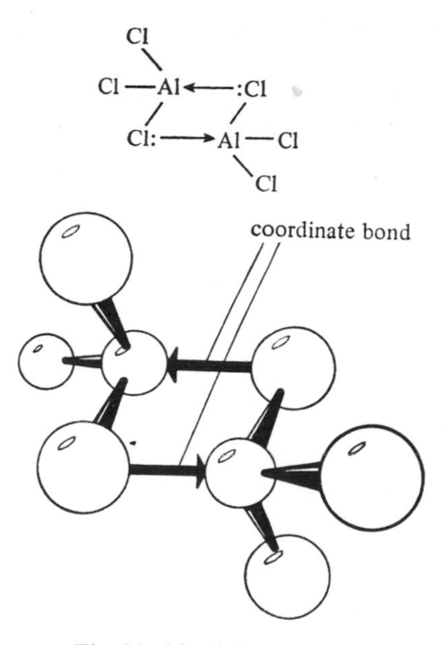

Fig. 21. Aluminium chloride

The electron configuration of aluminium is (Ne)$3s^2 3p^1$ and since aluminium is generally trivalent we may assume promotion of one of the $3s$ electrons to a vacant $3p$ orbital giving, (Ne)$3s^1 3p^1 3p^1$. To explain the equivalence of the three covalent bonds, in compounds like aluminium chloride the formation of sp^2 hybrid orbitals to give (Ne)$(3sp^2)^1(3sp^2)^1(3sp^2)^1$ is assumed. The sp^2 orbitals form directional bonds in a plane at an angle of 120 degrees. Electron pairing and sharing between one aluminium atom and three chlorine atoms, with sp^2–p overlap, predicts a covalent molecule of formula $AlCl_3$. Molecular weight determinations suggest that the aluminium chloride molecule is dimerized, (Al_2Cl_6) in the vapour state and this can be accounted for by supposing that an unshared pair of electrons, from the chlorine atom of one aluminium chloride molecule, can overlap into the empty *p* orbital of the aluminium atom of the other.

The pair of electrons provided for coordination by a chlorine atom of one molecule repels the bond pairs of the acceptor molecule and distorts the planar structure (Fig. 21). This type of chemical bond is similar to the covalent bond in that two electrons are shared between two overlapping orbitals, but differs in that both electrons are provided from an unshared electron pair of one atom. This kind of orbital overlap is called a *coordinate bond*, and may be represented by an arrow between the donor and acceptor atoms, involved in the sharing process.

Comparison of the properties of the aluminium halides (Table 15) reveals that aluminium fluoride may well be ionic, as would be predicted by Fajans' rules. Pauling has suggested that a giant molecular structure for aluminium fluoride would give an alternative explanation.

Table 15. Properties of aluminium halides

Property	*Fluoride*	*Chloride*	*Bromide*	*Iodide*
Melting point (K)	1313	463	370	464
Boiling point (K)	—	Sublimes	538	659
Water solubility	Slightly soluble	Hydrolyzed	Hydrolyzed	Hydrolyzed
Covalent solubility	Insoluble	Soluble	Soluble	Soluble

5.1 The reaction between ammonia and hydrogen chloride

The idea of coordinate bonds has already been touched upon at various points in Chapters 2 and 3, where it was suggested that molecules with unshared pairs of electrons could donate these to suitable acceptors to produce either temporary or permanent bonds.

It will be recalled that the ammonia molecule has such a projecting, unshared pair of electrons, and these can be used to form coordinate bonds with electron acceptors. When a molecule of ammonia approaches one of hydrogen chloride, the dipoles interact so that the unshared pair on the nitrogen atom coordinates onto the hydrogen of the hydrogen chloride, the bond between the hydrogen atom and the chlorine atom breaking to leave a

$$
{}^{\delta+}\mathrm{H}{-}\overset{\mathrm{H}\diagdown}{\underset{\mathrm{H}\diagup}{\mathrm{N}}}{:}\,{}^{\delta-} \qquad {}^{\delta+}\mathrm{H}{-}\mathrm{Cl}^{\delta-} \longrightarrow \left[\mathrm{H}{-}\overset{\mathrm{H}\diagdown}{\underset{\mathrm{H}\diagup}{\mathrm{N}}}{:}\rightarrow \mathrm{H}\right]^{+} \quad \mathrm{Cl}^{-}
$$

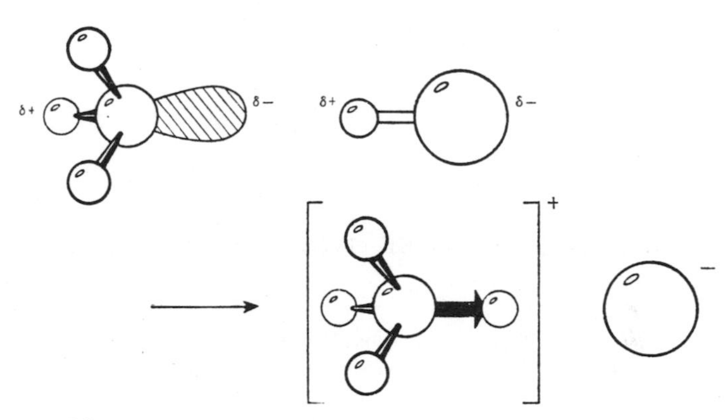

Fig. 22. Reaction between ammonia and hydrogen chloride

$$
{}^{\delta+}\mathrm{H}{-}\overset{\mathrm{H}\diagdown}{\underset{\mathrm{H}\diagup}{\mathrm{N}}}{:}\,{}^{\delta-} \qquad {}^{\delta+}\mathrm{H}{-}\overset{\delta-}{\mathrm{O}}\diagdown_{\mathrm{H}} \longrightarrow \left[\mathrm{H}{-}\overset{\mathrm{H}\diagdown}{\underset{\mathrm{H}\diagup}{\mathrm{N}}}{:}\rightarrow \mathrm{H}\right]^{+} \quad \mathrm{OH}^{-}
$$

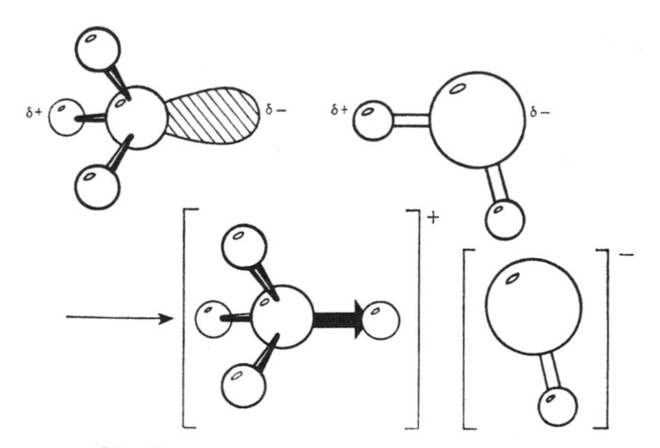

Fig. 23. Reaction between ammonia and water

chloride ion and an ammonium ion (Fig. 22). The ammonium ions and the chloride ions are mutually attracted to form an ionic lattice. The reader may care to speculate why ammonium chloride despite having a giant ionic lattice is apparently volatile.

5.2 The reaction between ammonia and water

The reaction between ammonia and water to form ammonium ions and hydroxyl ions also results from dipole–dipole attractions leading to the formation of a coordinate bond (Fig. 23).

5.3 The reaction between hydrogen chloride and water

The water molecule, with two unshared pairs of electrons, can also co-ordinate onto a proton to form a hydronium ion. For instance, when gaseous, covalent hydrogen chloride dissolves in water, dipole–dipole attraction followed by coordinate bond formation again occurs (Fig. 24).

$$\overset{\delta+}{H_2}\ddot{O}:^{\delta-} \quad {}^{\delta+}H-Cl^{\delta-} \longrightarrow \left[H_2\ddot{O}: \rightarrow H\right]^{+} \quad Cl^{-}$$

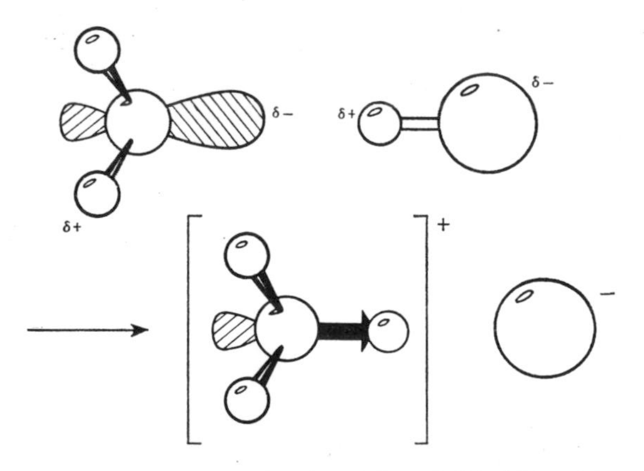

Fig. 24. Reaction between hydrogen chloride and water

5.4 Acid-Base Character

In the reaction between hydrogen chloride and water, the hydrogen chloride molecule has effectively donated a proton to the water molecule. Such a proton transfer is an example of an acid-base reaction if the system suggested by Brönsted and Lowry in 1923 is adopted.

They proposed that an acid be defined as a *proton donor* and a base as a *proton acceptor*. Hence, in the above example the hydrogen chloride molecule is behaving as an acid and the water molecule as a base. However the ionisation is reversible, and in the reverse reaction the hydronium ion donates a proton and the chloride ion accepts it.

$$\underset{\text{Acid}}{HCl} + \underset{\text{Base}}{H_2O} \rightleftharpoons \underset{\text{Acid}}{H_3O^+} + \underset{\text{Base}}{Cl^-}$$

The ions and molecules in this equilibrium comprise two acid-base pairs, and such pairs, which may be simply interconverted by the transfer of a proton are termed *conjugate acid-base pairs*. For example

$$\underset{\text{Acid}}{HCl} \underset{+H^+}{\overset{-H^+}{\rightleftharpoons}} \underset{\text{Base}}{Cl^-}$$

and

$$\underset{\text{Base}}{H_2O} \underset{-H^+}{\overset{+H^+}{\rightleftharpoons}} \underset{\text{Acid}}{H_3O^+}$$

The reaction between ammonia and water can be represented in a similar way:

$$\underset{\text{Base}}{NH_3} + \underset{\text{Acid}}{H_2O} \rightleftharpoons \underset{\text{Acid}}{NH_4^+} + \underset{\text{Base}}{OH^-}$$

The conjugate acid-base pairs in this case are

$$\underset{\text{Base}}{NH_3} \underset{-H^+}{\overset{+H^+}{\rightleftharpoons}} \underset{\text{Acid}}{NH_4^+}$$

$$\underset{\text{Acid}}{H_2O} \underset{+H^+}{\overset{-H^+}{\rightleftharpoons}} \underset{\text{Base}}{OH^-}$$

In this reaction water behaves as an acid. Substances which can act as both acids and bases are termed *amphoteric*.

A further and more generalized idea of acids was put forward in 1923 by Lewis. On this approach an acid is defined as an *electron pair acceptor*, and a base as an *electron pair donor*. This is relevant to the idea of the coordinate bond, in the formation of which electron pairs are accepted and donated. In Fig. 24 it can be seen that an electron pair from a water molecule is donated to the hydrogen chloride molecule prior to ionization. Therefore, water is a Lewis base and hydrogen chloride a Lewis acid.

Similarly in Fig. 23 the ammonia molecule donates an electron pair to a water molecule and ionization then follows. The ammonia is thus the Lewis base and water the Lewis acid; a further indication of the amphoteric nature of water.

5.5 The limited ionization of water

In order to account for the low but measurable conductivity of pure water and for the products obtained at the electrodes during the electrolysis of aqueous solutions limited ionization of water is frequently postulated. The formation of a coordinate bond between one water molecule and the hydrogen atom from another gives an explanation of this process. We can estimate the proportion of water molecules which are ionized. From measurements of the conductivity of water the hydronium ion concentration is 10^{-7} mol dm^{-3}. The Avogadro constant is 6×10^{23}.

$$[H_3O^+] = 10^{-7} \text{ mol dm}^{-3}$$

$$\text{the number of hydronium ions in every cubic decimetre} = 10^{-7} \times 6 \times 10^{23}$$

$$= 6 \times 10^{16}$$

$$\text{number of water molecules in every cubic decimetre} = \frac{1000}{18} \times 6 \times 10^{23}$$

$$= 3 \times 10^{25}$$

The ratio of hydronium ions to water molecules is therefore about 1 to 5×10^8, or, to put it another way, only one water molecule in 500000000 produces a hydronium ion.

However although the proportion of hydronium ions is low it should be appreciated that there are about 6×10^{16} hydronium ions in each cubic decimetre of water.

It will be seen from Fig. 25 that this reaction may also be regarded as an acid-base reaction by applying either the Brönsted and Lowry or the Lewis definition of an acid. The amphoteric nature of water is again made evident.

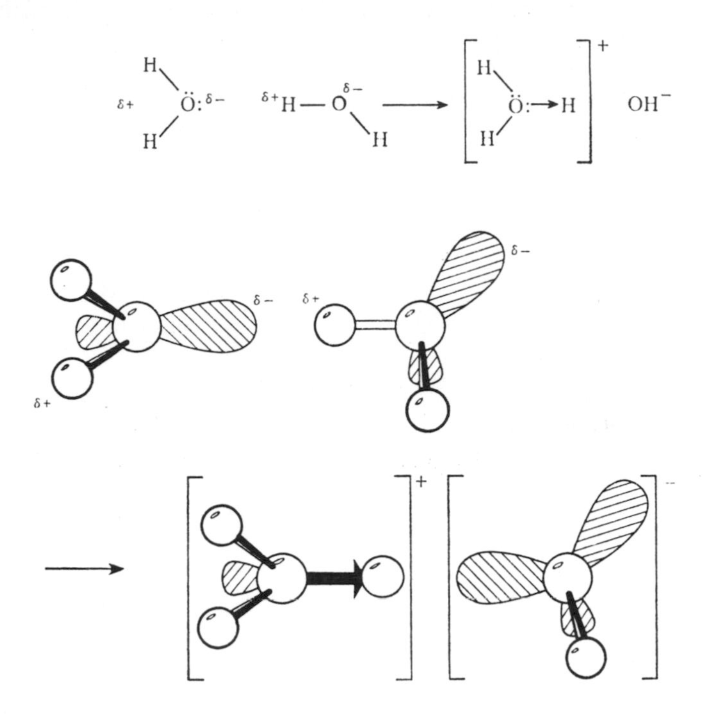

Fig. 25. Limited ionization of water

5.6 Some sulphur compounds

The coordinate bond is helpful in explaining the formation and structure of the oxides and oxy-acids of sulphur. Using an arrow to represent the co-ordinate bond, Fig. 26 illustrates the relationship between these compounds.

The formation of the coordinate bond may be visualized as follows. In sulphur dioxide, for example, the sulphur atom, $(Ne)3s^2 3p^2 3p^1 3p^1$, shares and pairs the two $3p$ electrons with the two $2p$ electrons of an oxygen atom, $(He)2s^2 2p^2 2p^1 2p^1$, forming two covalent bonds between the two atoms. The second oxygen atom internally pairs off the two $2p^1$ electrons, leaving a vacant $2p$ orbital for coordinate overlap involving the unshared $3p^2$ electrons from the sulphur. Since sulphur dioxide on conversion to sulphur

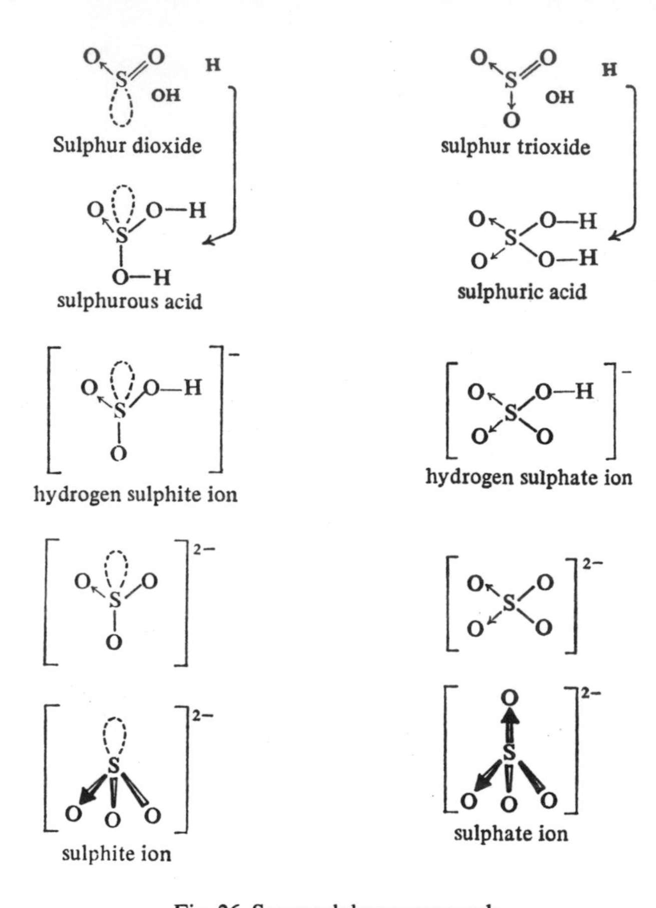

Fig. 26. Some sulphur compounds

trioxide appears to use a second unshared pair in forming a coordinate bond with another oxygen atom, some prior formation of hybrid orbitals between the *s* and the *p* orbitals seems to be necessary.

When sodium sulphite is boiled with sulphur, sodium thiosulphate is formed, the sulphite ion incorporating a sulphur atom to form the thiosulphate ion. Acidification gives the parent acid, thiosulphuric acid, which is unstable and decomposes to give sulphur and sulphur dioxide. These changes are represented in Fig. 27.

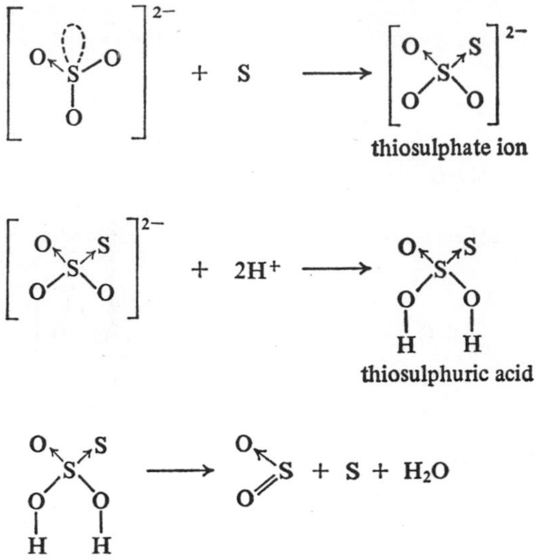

thiosulphate ion

thiosulphuric acid

$$\text{thiosulphuric acid} \longrightarrow SO_2 + S + H_2O$$

Fig. 27. Thiosulphate ion

Sulphamic acid is similar to sulphuric acid with the amino group (NH_2) replacing one of the OH groups (Fig. 28).

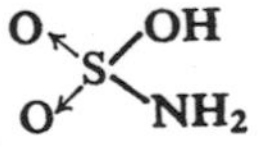

Fig. 28. Sulphamic acid

5.7 Some oxy-nitrogen compounds

Figure 29 shows the relationship between nitrous and nitric acids and their anhydrides. The unshared pair of electrons on the nitrite ion by coordination onto available oxygen atoms may account for the reducing properties.

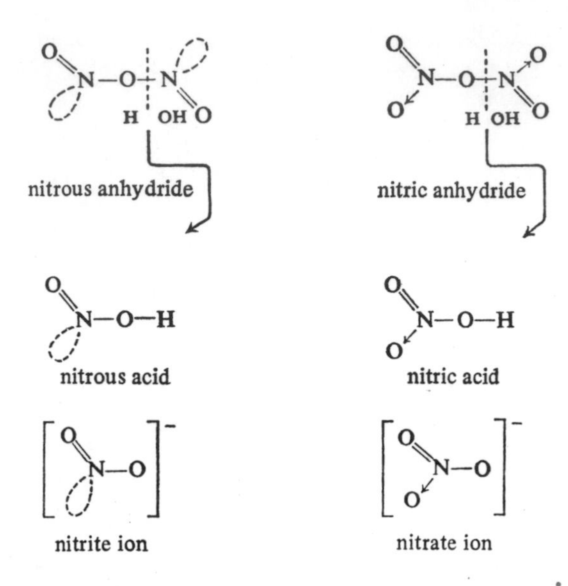

Fig. 29. Some oxy-nitrogen compounds

5.8 Some oxy-phosphorus compounds

The structures of the oxides and oxy-acids of phosphorus can also be explained in terms of the coordinate bond, as shown in Fig. 30.

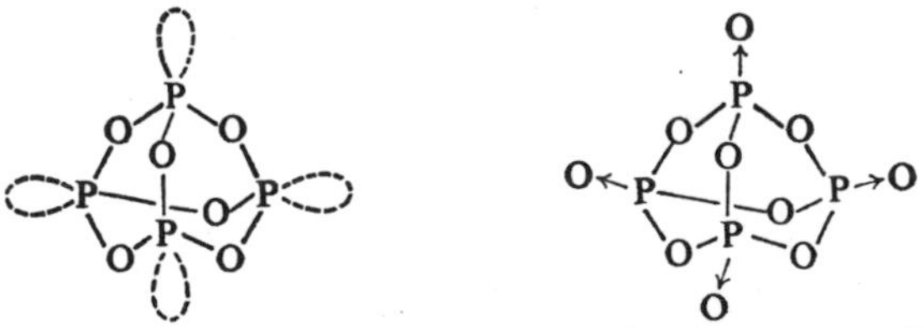

phosphorus trioxide
hydrolysis gives

phosphorus pentoxide
partial hydrolysis of P–O–P bonds gives

Fig. 30 (*continued on next page*)

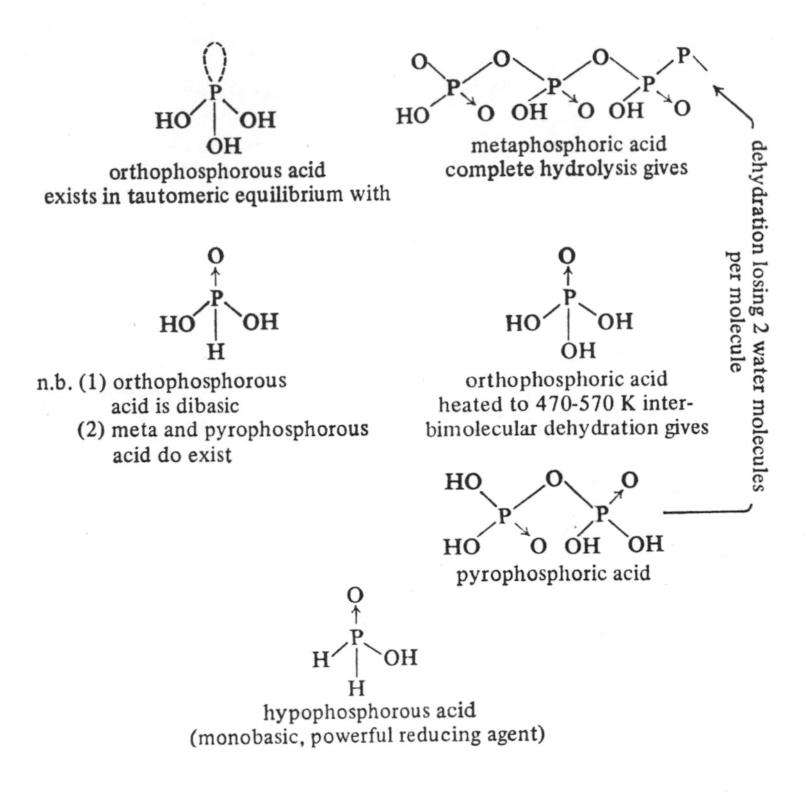

Fig. 30. Some oxy-phosphorus compounds

5.9 Oxy-acids of chlorine

When chlorine is dissolved in water, reaction occurs to give hypochlorous acid and hydrochloric acid, $Cl_2 + 2H_2O \rightleftharpoons H_3O^+ + Cl^- + HOCl$.

The structure of the hypochlorous acid molecule (Fig. 31) shows three unshared pairs of electrons on the chlorine atom and these, on successive coordination onto three oxygen atoms, give plausible structures for chlorous acid, chloric acid and perchloric acid.

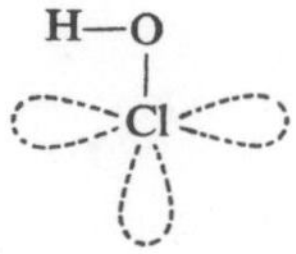

H—O
|
Cl
hypochlorous acid

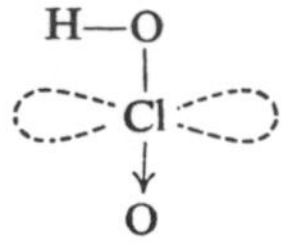

H—O
|
Cl
↓
O
chlorous acid

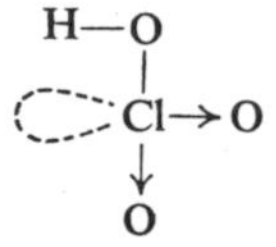

H—O
|
Cl→O
↓
O
chloric acid

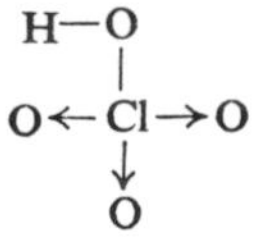

H—O
|
O←Cl→O
↓
O
perchloric acid

Fig. 31. Oxy-acids of chlorine

5.10 Limitations of models

This might be a good time to remember that we are using models to represent structures in an attempt to understand the behaviour of substances. We are not saying that the models are exact.

For instance, in the sulphate ion (Fig. 26), there are two covalent and two coordinate bonds and this may give the impression that the ion has bonds of different type and possibly of different length. Measurements show that the four bonds are identical.

Further the drawings infer that the molecules are planar. The reader could attempt to work out the probable arrangement in space of the atoms for some of the molecules discussed in this chapter, either by building models or by using ideas of repulsion between bond pairs and lone pairs.

6 Transition Elements

Reference to the energy level diagram, Fig. 1, shows that up to the element argon, the levels to be filled follow successive numerical order, 1*s*, 2*s*, 2*p*, 3*s* and 3*p*. The level 3*d*, containing five orbitals, is higher than 4*s* and consequently, after calcium of configuration $(Ar)4s^2$, ten elements can be accommodated as the 3*d* level fills from $3d^1$ to $3d^{10}$. These elements, scandium (Sc) to zinc (Zn), see Table 16, make up the first series of transition elements. The situation is repeated after krypton, when rubidium and strontium show the filling of the 5*s* level before the 4*d* level fills to give a second series of transition elements, from yttrium (Y) to cadmium (Cd). Again after xenon the 6*s* level fills, with caesium (Cs) and barium (Ba), in advance of the 5*d* level, but the 4*f* level, containing seven orbitals and accounting for fourteen elements, is above 6*s* and below 5*d*. As a result, there are twenty-four elements from lanthanum to mercury, ten transition elements, in the same sense as above, and a set of fourteen inner transition elements, the lanthanides.

The close resemblance between the electron configurations of the transition elements, a more or less complete penultimate *d* level, with one or two electrons in the valency shell, causes them to be similar in character, but the reactivity varies considerably, generally falling off from the lightest to the heaviest in any series.

None of the elements can attain a noble gas configuration by electron gain, loss or sharing. The energy needed to remove one or two electrons is of the same order of magnitude as for the alkali metals and the alkaline earths (compare Table 11 with Table 16). Lattice energies for ionic compounds formed from transition elements are similar to those of alkaline earths and, in these terms, it is not surprising that transition elements can, and do, react by electron loss to electron acceptors, forming positive ions. With unpaired electrons in the structures a measure of covalence is to be expected and occurs.

6.1 Variable valency

The presence in transition elements of partially filled *d*-levels leads to a wide variety of bonding possibilities. It is by using these possibilities in different ways that transition elements are able to appear in several valency states. All valencies from one to eight are known, and network or layer lattice structures, intermediate between ionic and covalent in bonding, are common (Table 17). The variety of oxidation states, shown in Table 16, is another illustration of the alternative modes of electron arrangement available to transition elements. The extent of variation is greatest towards the centre of a series, where the number of electrons of unpaired spin reaches a maximum.

Table 16. Properties of the elements of first transition series

Element	*Electron configuration*	*Melting point* (K)	*Boiling point* (K)	*Density* (g cm^{-3})	*Heat of atomization* (kJ mol^{-1})	*ionization energies* (kJ mol^{-1})			*Oxidation states* (+)
						1st	2nd	3rd	
Sc	(Ar)$3d^1 4s^2$	1813	3003	3·0	343	632	1866	4268	3
Ti	(Ar)$3d^1 3d^1 4s^2$	1943	3533	4·5	473	657	1966	4644	2, 3, 4
V	(Ar)$3d^1 3d^1 3d^1 4s^2$	2173	3723	6·1	515	649	2063	4602	2, 3, 4, 5
Cr	(Ar)$3d^1 3d^1 3d^1 3d^1 3d^1 4s^1$	2143	2933	7·1	398	653	2243	5314	2, 3, 6
Mn	(Ar)$3d^1 3d^1 3d^1 3d^1 3d^1 4s^2$	1523	2423	7·4	279	716	2226	5523	1, 2, 3, 4, 6, 7
Fe	(Ar)$3d^2 3d^1 3d^1 3d^1 3d^1 4s^2$	1813	3273	7·9	418	757	2318	5230	2, 3, 6
Co	(Ar)$3d^2 3d^2 3d^1 3d^1 3d^1 4s^2$	1763	3173	8·9	427	757	2402	5690	2, 3, 4
Ni	(Ar)$3d^2 3d^2 3d^2 3d^1 3d^1 4s^2$	1723	3003	8·9	423	736	2485	5941	2, 3, 4
Cu	(Ar)$3d^2 3d^2 3d^2 3d^2 3d^2 4s^1$	1356	2873	9·0	339	745	2703	6360	1, 2, 3
Zn	(Ar)$3d^2 3d^2 3d^2 3d^2 3d^2 4s^2$	692	1179	7·1	131	908	2640	6485	2

Cr and Cu are the only elements with a $4s^1$ outer orbital. The configuration with a half-filled and completely filled *d* orbital is particularly stable.

Table 17. Chlorides and oxides of transition elements

Compounds	*Formula*	*Valency*	*Structure*
Titanium(IV) chloride	$TiCl_4$	4	Covalent
Iron(II) chloride	$FeCl_2$	2	Layer lattice
Iron(III) chloride	$FeCl_3$	3	Layer lattice
Cobalt(II) chloride	$CoCl_2$	2	
Copper(I) chloride	$CuCl$	1	Network lattice
Copper(II) chloride	$CuCl_2$	2	Chain lattice
Silver(I) chloride	$AgCl$	1	Network lattice
Titanium(IV) oxide	TiO_2	4	Network lattice
Vanadium(V) oxide	V_2O_5	5	Layer lattice
Chromium(VI) oxide	CrO_3	6	Chain lattice
Manganese(IV) oxide	MnO_2	4	Non-stoichiometric
Manganese(VII) oxide	Mn_2O_7	7	Covalent
Iron(II) oxide	FeO	2	Network lattice
Iron(III) oxide	Fe_2O_3	3	Layer lattice
Cobalt(II) oxide	CoO	2	Network lattice
Copper(I) oxide	Cu_2O	1	Network lattice
Copper(II) oxide	CuO	2	
Osmium(VIII) oxide	OsO_4	8	Covalent

6.2 Colour

In contrast to the alkali metals, alkaline earths and group III metals, transition elements generally form coloured compounds. Some of these are summarized in Table 18.

Table 18. Colours of compounds of first transition series

Element	*No. of unpaired electrons*	*Oxidation state* +2	+3	+4	+5	+6	+7
Sc	1		Colourless				
Ti	2		Violet	Colourless			
V	3	Violet	Green	Blue	Red		
Cr	6	Blue	Green			Orange	
Mn	5	Pink				Green	Purple
Fe	4	Green	Yellow				
Co	3	Pink	Blue				
Ni	2	Green					
Cu	1	Blue					
Zn	0	Colourless					

When white light falls on an array of atoms, molecules or ions, energy can be transferred to the particles only if the energy associated with a particular wavelength in the visible region of the electromagnetic spectrum, corresponds exactly with the energy required to promote an electron to a higher energy level. The wavelength is then absorbed from the incident light; the emergent light thus has a component missing and appears coloured.

For the ions of the elements of groups I, II and III the energy difference between the orbital occupied by the outermost electrons and the next available level is considerable, and cannot be attained by absorption of light energy with frequencies in the visible region. The ions are in consequence, colourless. For transition elements on the other hand, the energy required to raise *d* electrons to other *d* orbitals of higher energy with suitable vacancies is of the right order. Hence compounds of transition elements are frequently coloured.

The *d* orbitals in an isolated atom or ion are equivalent in energy, they are said to be *degenerate*, but they do not remain so in the presence of approaching atoms, molecules or ions which can donate an electron pair to form a coordinate bond. Such an electron donor is referred to as a *ligand*, and tends to repel the electrons in the *d* orbitals closest to it more than it does the electrons in other *d* orbitals. The energies of the *d* orbitals thus become unequal, and it is the transitions of electrons between these different energy levels which are responsible for the absorption of electromagnetic radiation from the visible region.

For example, copper(II) ions in an aqueous environment absorb the red colour of visible light and the light transmitted by the ions is the complementary colour blue. In the case of the copper(II) ion in anhydrous copper sulphate however, the *d* orbitals are all equivalent since the ion is not coordinated onto molecules of water, and hence the *d* orbitals are equivalent in energy and there are no orbital transitions of sufficiently low energy available. Anhydrous copper sulphate therefore appears white.

The colour of a transition metal compound will be influenced by the nature of the ligands surrounding the central atom. A second factor influencing the colour of such compounds is the oxidation state of the central atom. This will affect the number of orbital vacancies available and hence the transitions which are possible.

6.3 Paramagnetism

When a *paramagnetic* substance is placed in a magnetic field the constituent atoms tend to line up under the influence of the field. Substances which are not subject to this influence are said to be *diamagnetic*. Paramagnetic substances are attracted by a magnetic field, diamagnetic substances are not.

Many transition elements and their compounds are paramagnetic and this phenomenon is associated with electrons of unpaired spin. A spinning electron causes weak electrical and magnetic fields, and where an orbital is paired the equal and opposite spins produce equal and opposite fields which cancel, to give a zero resultant field. An unpaired electron causes a small, but significant, resultant field which interacts with the external magnetic field. The greater the number of unpaired electrons, the greater the paramagnetic effect, as illustrated in Fig. 32.

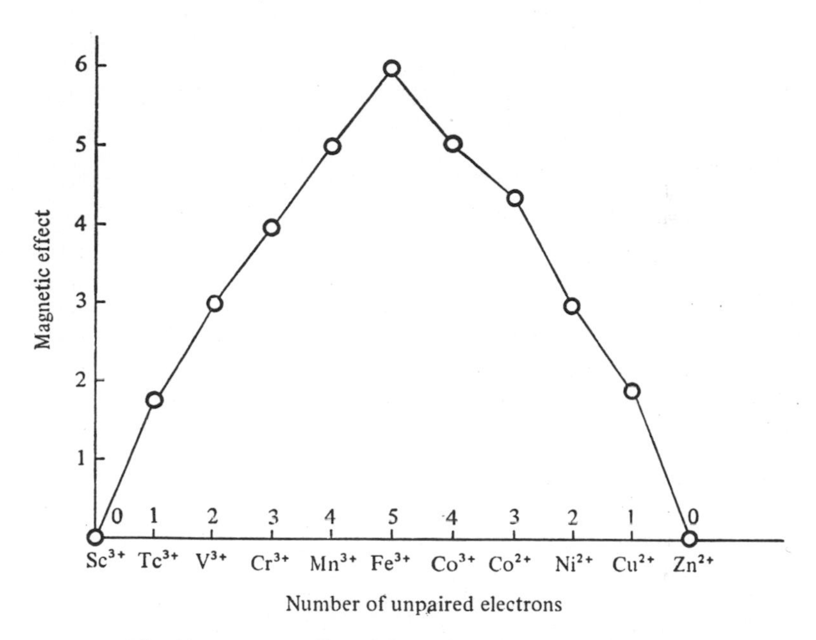

Fig. 32. Magnetic effect and number of unpaired electrons

The paramagnetism of some transition metal compounds is difficult to relate to the apparent number of unpaired electrons suggested by simple ideas of bonding. For instance, compounds containing the ion FeF_6^{3-} give rise to a high paramagnetic effect corresponding to five unpaired electrons, whereas others, such as those containing the $Fe(CN)_6^{3-}$ ion, have a smaller effect, corresponding to only one unpaired electron. Various explanations can be offered for this discrepancy and a simplified, account can be based on ideas already introduced in discussing the colour of transition metal compounds.

As indicated previously, the energies of the *d* orbitals of a transition metal atom or ion will be influenced by the surrounding ligands, whether these are negative ions or polar molecules. These ligands will repel the electrons in the five *d* orbitals to different extents resulting in a splitting of the previously equal energy levels. The extent of the change brought about by the surrounding ligands will depend upon their nature, and the way in which the new orbitals are occupied by the electrons is determined by the energy differences between them.

In the isolated Fe^{3+} ion the five electrons in the *d* orbitals are unpaired and the situation may be represented thus:

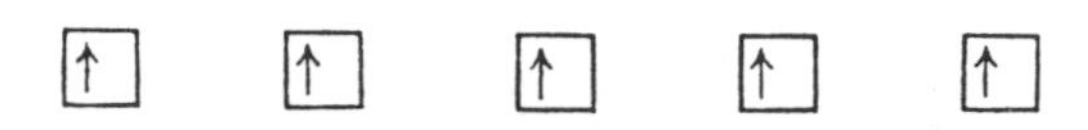

Five equivalent *d* orbitals

In the presence of the six F^- ligands the five orbitals are no longer equivalent, and split into two sets with different energies. The three orbitals with the lower energy are called d_ϵ and the two with the higher energy d_γ. The F^- ions produce only a weak field which means that the energy difference between the two sets of orbitals is not great, and the five electrons, despite the higher energy of the two d_γ orbitals, still remain unpaired. This is represented as follows.

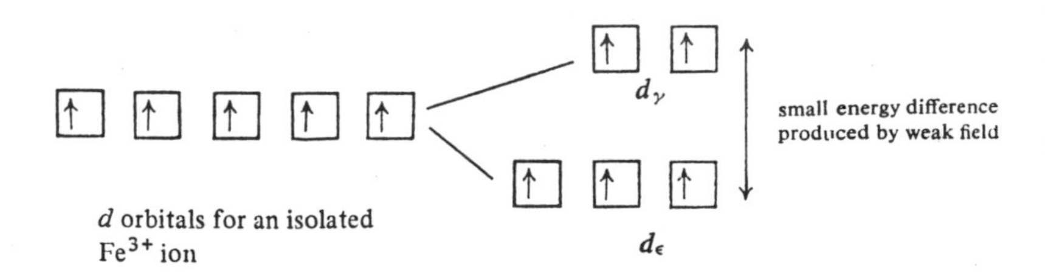

The formation of a coordinate bond requires a vacant orbital, and since in FeF_6^{3-} the five electrons remain unpaired the ligands must use other vacant orbitals. The paramagnetic effect therefore corresponds to five unpaired electrons.

For the case of the $Fe(CN)_6^{3-}$ ion, however, the field produced by the CN^- ions is greater, and the energy difference between the d_ϵ and d_γ sets of orbitals is larger. Under these conditions, the lowest energy for the complex is when as many electrons as possible occupy the lower set of orbitals, even though this means that the stability which follows from the single occupation of orbitals is sacrificed. This can be represented:

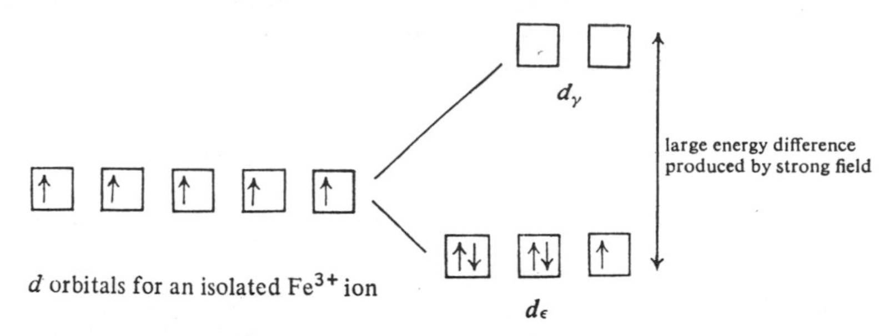

It can be seen that two of the *d* orbitals are now available for use by the coordinating ligands and only one electron is unpaired so explaining the low paramagnetic effect.

Extensions of this simplified treatment, termed *crystal field theory*, and the more sophisticated *ligand field theory* have been useful in explaining the magnetic properties shown by a variety of transition metal compounds.

Ferromagnetism is an extreme form of paramagnetism in which the substance retains a residual magnetic property in the absence of an external magnetic field. It is a much stronger effect than paramagnetism, and ferromagnetic substances attract one another, unlike those which are paramagnetic. For an element to be ferromagnetic it must have electrons of unpaired spin. The atoms must also be sufficiently far apart to prevent overlap of singly occupied orbitals on neighbouring atoms, which would cause pairing of opposed spins, but near enough for the lining up of electron spins in neighbouring atoms. That the interatomic distance is critical is illustrated by manganese, which despite five unpaired electrons is not ferromagnetic, though some of its compounds are. The close packing of atoms in the metal results in pairing the 3*d* electrons between neighbouring atoms. In a compound such as manganese nitride, the nitrogen atoms keep the manganese atoms far enough apart to prevent overlap.

6.4 Catalytic activity

Transition elements and their compounds are remarkable for their ability to catalyse a wide variety of chemical reactions (Table 19). Though the way catalysts work is only partially understood, it is believed that the presence of unpaired or unfilled *d* orbitals is an important factor. Intermediate complexes can form, utilizing the *d* orbitals and providing reaction routes of favourable energy, compared to the routes available without the catalyst.

Table 19. Examples of catalytic activity of transition elements

Catalyst	*Process*	*Equation*
Platinum	Contact process	$2SO_2 + O_2 \rightleftharpoons 2SO_3$
Platinum	Oxidation of ammonia	$4NH_3 + 5O_2 \rightleftharpoons 4NO + 6H_2O$
Iron	Haber process	$N_2 + 3H_2 \rightleftharpoons 2NH_3$
Iron/Chromium	Bosch reaction	$CO + H_2O \rightleftharpoons CO_2 + H_2$
Nickel	Hydrogenation of olefines	$H_2C{:}CH_2 + H_2 \rightleftharpoons H_3C.CH_3$
Cobalt	Manufacture of butadiene	$2H_2C{:}CH_2 \rightleftharpoons H_2C{:}CH.CH{:}CH_2 + H_2$
Copper	Manufacture of formaldehyde	$2CH_3OH + O_2 \rightleftharpoons 2HCHO + 2H_2O$
Vanadium pentoxide	Contact process	$2SO_2 + O_2 \rightleftharpoons 2SO_3$
Manganese dioxide	Decomposition of potassium chlorate	$2KClO_3 \rightarrow 2KCl + 3O_2$

7 Complexes

One of the characteristics of transition metal ions, although it is not confined to them, is their ability to form complexes. This property has already been referred to when dealing with colour, paramagnetism and catalytic behaviour but it is sufficiently important to merit a separate section.

Some complex ions carry a positive charge, some a negative charge, others are neutral. All are coordination compounds in which the central metal ion is surrounded by a number of coordinating ions or molecules. These ligands must have at least one unshared pair of electrons in a bonding orbital. The relative size of ligand and central ion is an important factor in determining the maximum number of ligands which can be accommodated. The *coordination number* of an ion is the number of coordinate bonds which it accepts in forming a complex. It should not be confused with the meaning of the term in simple ionic lattices, where it means the number of ions in close contact with a central ion. Some common ligands are summarized in Table 20.

Table 20. Common ligands

Ligand	*Name*	*Abbreviation*	*No. of unshared pairs utilized in complex formation*
F^-	Fluoride ion		1
Cl^-	Chloride ion		1
Br^-	Bromide ion		1
I^-	Iodide ion		1
OH^-	Hydroxyl ion		1
CN^-	Cyanide ion		1
H_2O	Water		1
NH_3	Ammonia		1
N	Pyridine	(py)	1
CO_3^{2-}	Carbonate ion		2
$C_2O_4^{2-}$	Oxalate ion	(ox)	2
$H_2N.CH_2.CH_2.NH_2$	Ethylenediamine	(en)	2
$H_2C.N(CH_2COOH)_2$ / | / $H_2C.N(CH_2COOH)_2$	Ethylenediamine-tetracetic acid	(EDTA)	6

7.1 Naming of complexes

1. All anionic ligands end in -o (for example chloro). Molecular ligands take their own name except for water (aquo) and ammonia (ammine).

2. Complexes are named as single words with the names of ligands preceding the name of the metal ion, the oxidation state of which is shown as a Roman numeral in parentheses. Anionic ligands are placed before neutral ligands, and alphabetical order is used, within either class.

3. For complex anions the suffix- ate is added to the name or stem of the metal ion.

Table 21. Examples of nomenclature of complexes

Formula	*Name*
$[Pt(NH_3)_4Cl_2]^{2+}$	Dichlorotetrammineplatinum(IV) ion
$[Co(NH_3)_2(NO_2)_2(CN)_2]^-$	Dicyanodinitrodiamminecobaltate(III) ion
$[Ni(NH_3)_6]^{2+}$	Hexamminenickel(II) ion
$[HgI_4]^{2-}$	Tetraiodomercurate(II) ion
$[Cu(NH_3)_4]\,[CuCl_4]$	Tetramminecopper(II) tetrachlorocuprate(II)

7.2 An explanation of complex formation

One theory accounting for the formation and shape of complexes, uses the idea of hybrid orbitals to make available the required number of equivalent, vacant orbitals for bond formation. Many complexes can be explained in terms of hybrid orbitals (Table 22). The shapes of each of the four types are shown in Fig. 33, and the mechanism of complex formation in Tables 23 to 26.

Table 22. Varieties of hybridization and shape of complexes

Coordination number	*Orbitals hybridized*		*Shape*	*Example*
2	One *s* and one *p*	sp	Linear	$[Ag(NH_3)_2]^+$
4	One *s* and three *p*	sp^3	Tetrahedral	$[Zn(NH_3)_4]^{2+}$
4	One *d*, one *s* and two *p*	dsp^2	Square coplanar	$[Pt(NH_3)_4]^{2+}$
6	Two *d*, one *s* and three *p*	d^2sp^3	Octahedral	$[Co(NH_3)_6]^{3+}$

Table 23. Diamminesilver(I) ion, coordination number 2

Configuration of neutral atom	Ag	$(Kr)4d^{10}5s^1$
Configuration of central ion	Ag^+	$(Kr)4d^{10}$
Two vacant orbitals needed	Ag^+	$(Kr)4d^{10}(5s)^0(5p)^0$
Orbitals hybridized	Ag^+	$(Kr)4d^{10}(sp)^0(sp)^0$
Two ammonia molecules coordinate into vacant hybrid orbitals	Ag^+	$(Kr)4d^{10}(sp)\,(sp)$ ↑ ↑ $\ddot{N}H_3\ddot{N}H_3$
sp hybrid orbitals have a linear distribution		$[H_3N{\rightarrow}Ag{\leftarrow}NH_3]^+$

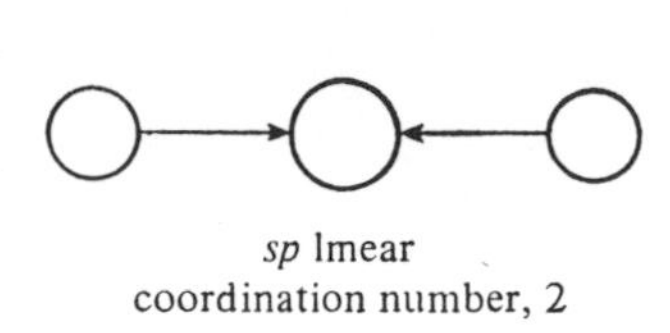

sp lmear
coordination number, 2

dsp^2 square coplanar
coordination number, 4

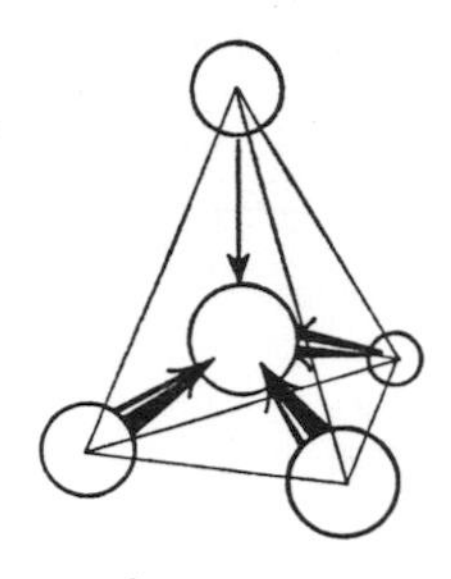

sp^3 tetrahedral
coordination number, 4

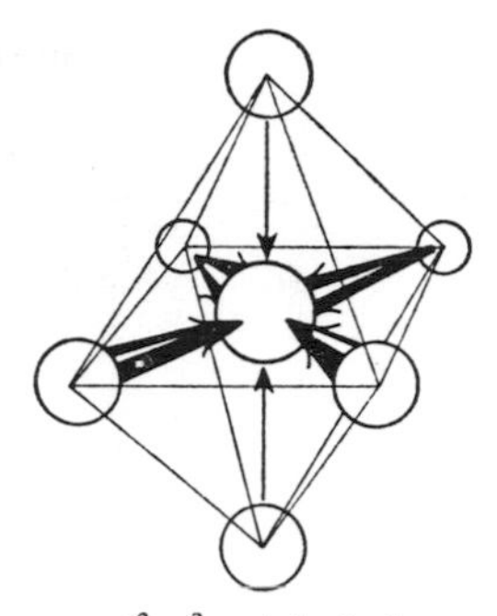

d^2sp^3 octahedral
coordination number, 6

Fig. 33. Shape of four varieties of complex

Table 24. Tetramminezinc(II) ion, coordination number 4

Configuration of neutral atom	Zn	$(Ar)3d^{10}4s^2$
Configuration of central ion	Zn^{2+}	$(Ar)3d^{10}$
Four vacant orbitals needed	Zn^{2+}	$(Ar)3d^{10}(4s)^0(4p)^0(4p)^0(4p)^0$
Orbitals hybridized	Zn^{2+}	$(Ar)3d^{10}(sp^3)^0(sp^3)^0(sp^3)^0(sp^3)^0$
Four ammonia molecules coordinate into vacant hybrid orbitals	Zn^{2+}	$(Ar)3d^{10}(sp^3)\ (sp^3)\ (sp^3)\ (sp^3)$ $\uparrow \quad \uparrow \quad \uparrow \quad \uparrow$ $\ddot{N}H_3\ \ddot{N}H_3\ \ddot{N}H_3\ \ddot{N}H_3$

sp^3 hybrid orbitals have a tetrahedral distribution

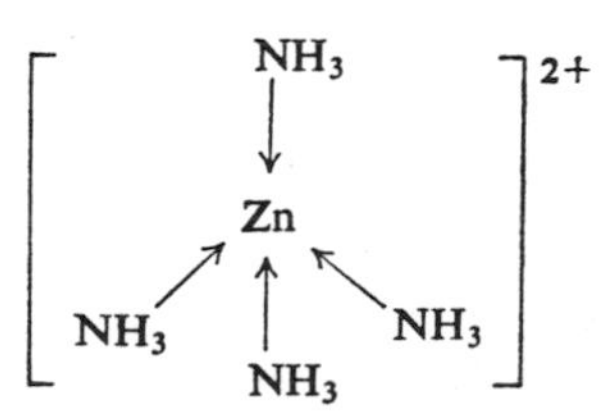

Table 26. Hexamminecobalt(III) ion, coordination number 6

Configuration of neutral atom	Co	$(Ar)3d^2 3d^2 3d^1 3d^1 3d^1 4s^2$
Configuration of central ion	Co^{3+}	$(Ar)3d^2 3d^1 3d^1 3d^1 3d^1$
Six vacant orbitals needed, two of the 3*d* orbitals paired	Co^{3+}	$(Ar)3d^2 3d^2 3d^2 (3d)^0 (3d)^0 (4s)^0 (4p)^0 (4p)^0 (4p)^0$
Orbitals hybridized	Co^{3+}	$(Ar)3d^2 3d^2 3d^2 (d^2sp^3)^0 (d^2sp^3)^0 (d^2sp^3)^0 (d^2sp^3)^0 (d^2sp^3)^0 (d^2sp^3)^0$
Six ammonia molecules coordinate into vacant hybrid orbitals	Co^{3+}	$(Ar)3d^2 3d^2 3d^2 (d^2sp^3)(d^2sp^3)(d^2sp^3)(d^2sp^3)(d^2sp^3)(d^2sp^3)$ ↑ ↑ ↑ ↑ ↑ ↑ $\ddot{N}H_3$ $\ddot{N}H_3$ $\ddot{N}H_3$ $\ddot{N}H_3$ $\ddot{N}H_3$ $\ddot{N}H_3$

d^2sp^3 hybrid orbitals have an octahedral distribution

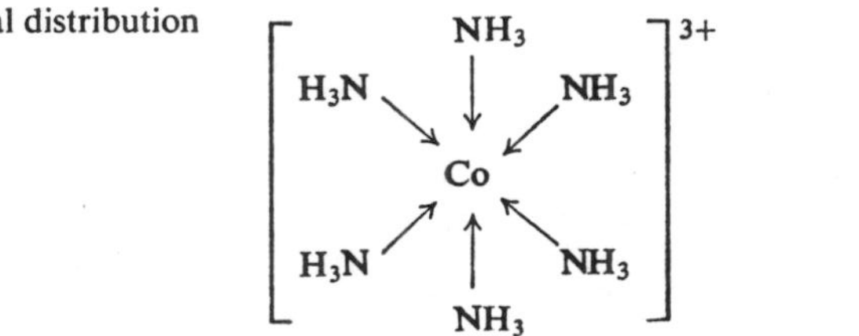

Table 25. Tetrammineplatinum(II) ion, coordination number 4

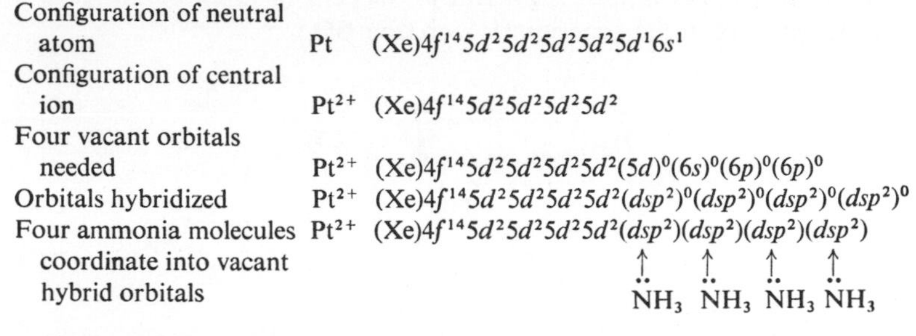

Configuration of neutral atom	Pt	$(Xe)4f^{14}5d^2 5d^2 5d^2 5d^2 5d^1 6s^1$
Configuration of central ion	Pt^{2+}	$(Xe)4f^{14}5d^2 5d^2 5d^2 5d^2$
Four vacant orbitals needed	Pt^{2+}	$(Xe)4f^{14}5d^2 5d^2 5d^2 5d^2 (5d)^0 (6s)^0 (6p)^0 (6p)^0$
Orbitals hybridized	Pt^{2+}	$(Xe)4f^{14}5d^2 5d^2 5d^2 5d^2 (dsp^2)^0 (dsp^2)^0 (dsp^2)^0 (dsp^2)^0$
Four ammonia molecules coordinate into vacant hybrid orbitals	Pt^{2+}	$(Xe)4f^{14}5d^2 5d^2 5d^2 5d^2 (dsp^2)(dsp^2)(dsp^2)(dsp^2)$ ↑ ↑ ↑ ↑ $\ddot{N}H_3$ $\ddot{N}H_3$ $\ddot{N}H_3$ $\ddot{N}H_3$

dsp^2 orbitals have a square coplanar distribution

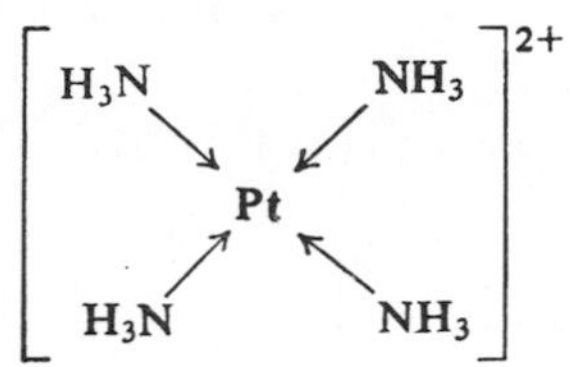

Some familiar complexes, and the type to which they belong are summarized in Table 27.

Table 27. Examples of types of complex

Type	*Hydrates*	*Cyanides*	*Ammines*	*Halides*
sp	$[Ag(H_2O)_2]^+$	$[Ag(CN)_2]^-$	$[Ag(NH_3)_2]^+$	$[AgCl_2]^-$
sp^3		$[Zn(CN)_4]^{2-}$	$[Cu(NH_3)_4]^+$	$[AlBr_4]^-$
dsp^2	$[Cu(H_2O)_4]^{2+}$	$[Ni(CN)_4]^{2-}$	$[Cu(NH_3)_4]^{2+}$	$[CuCl_4]^{2-}$
d^2sp^3	$[Co(H_2O)_6]^{2+}$	$[Fe(CN)_6]^{3-}$	$[Cr(NH_3)_6]^{3+}$	$[AlF_6]^{3-}$

7.3 Chelates

Ligands with more than one unshared pair of electrons available for coordination can, in suitable circumstances, form stable complexes involving ring closure. This process is termed chelation. Five- or six-membered rings are relatively strain free, and chelates from ligands with unshared pairs in the 1,4 or 1,5 positions are common.

Chelates are important in analytical chemistry, as illustrated by the identification and estimation of nickel. When nickel(II) ions, in aqueous solution, are treated with a solution of dimethyl glyoxal in ethanol, a red precipitate of dimethylglyoximononickel(II) is formed (Fig. 34).

Chlorophyll, essential for the conversion of carbon dioxide to carbohydrates in photosynthesis, is present in the cells of green plants. It is a chelate with a central magnesium(II) ion (Fig. 35).

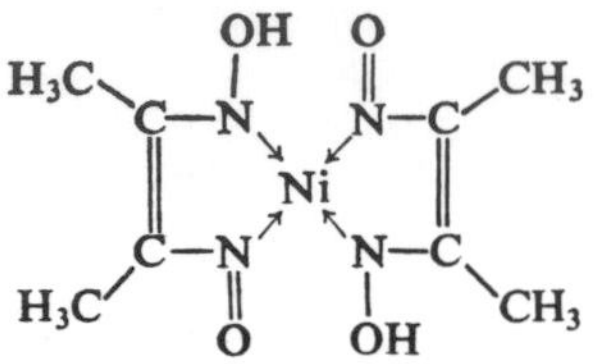

Fig. 34. Dimethylglyoximononickel(II)

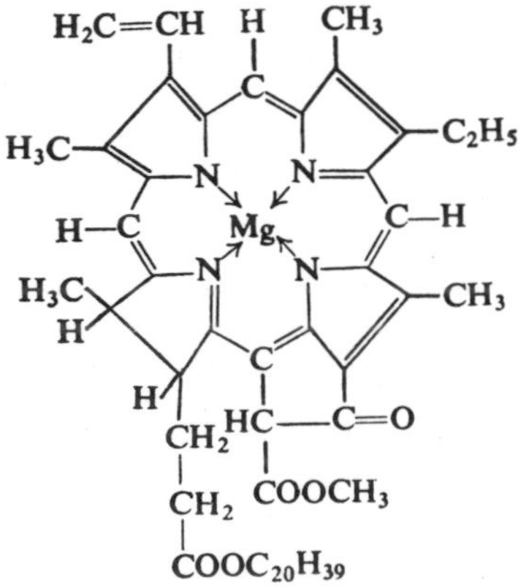

Fig. 35. Chlorophyll

The biologically active component of haemoglobin, responsible for oxygen transport in the blood stream, is a chelate with a central iron(III) ion (Fig. 36).

7.4 Complexones

Complexones are a class of aminopolycarboxylic acids, which have proved useful in the volumetric analysis of many elements.

The best known of them is ethylenediaminetetracetic acid (EDTA) which, when ionized, has four oxygen atoms capable of forming coordinate bonds, not to mention the unshared pair of electrons on each nitrogen atom giving up to six possible points of attachment for coordination on to a metal ion (Fig. 37). An important practical application of EDTA is the volumetric estimation of magnesium and calcium ions in aqueous solution. The disodium salt of EDTA is preferred to the free acid, which is not particularly soluble in water. The EDTA reagent, on adding to a solution containing the metal ions, forms a stable complex, effectively removing the metal ions from solution but remaining dissolved.

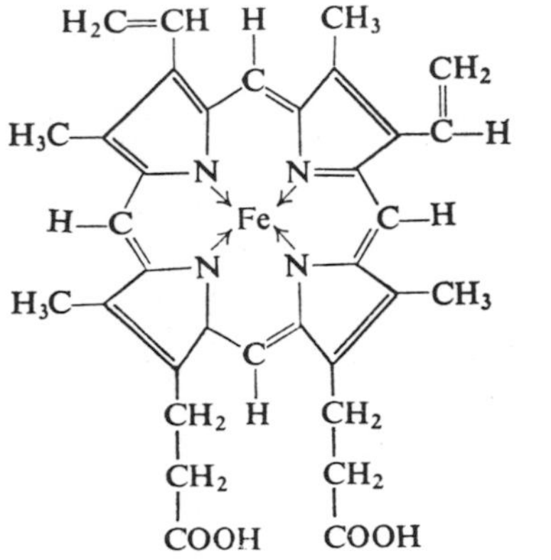

Fig. 36. Active component of haemoglobin

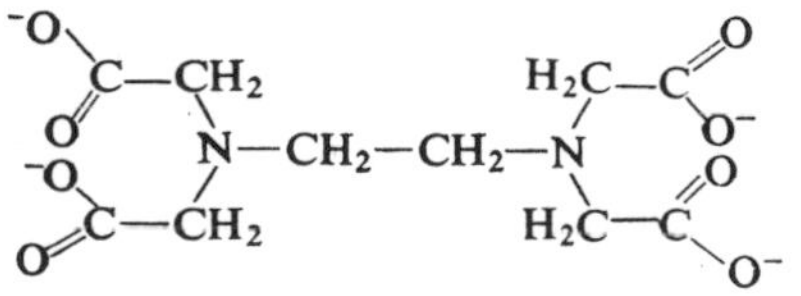

Fig. 37. EDTA ion

7.5 Isomerism

The directional nature of the ligand–ion bond is responsible for the existence of stereoisomers, geometrical and optical isomers being known.

Geometrical (cis-trans) isomerism

The complex neutral molecule dichlorodiammineplatinum(II) can be synthesized in two ways:

$$[PtCl_4]^{2-}_{(aq)} + 2NH_{3(aq)} \rightarrow Pt(NH_3)_2Cl_{2(s)} + 3Cl^-_{(aq)}$$

and

$$[Pt(NH_3)_4]^{2+}_{(aq)} + 2Cl^-_{(aq)} \rightarrow Pt(NH_3)_2Cl_{2(s)} + 2NH_{3(aq)}$$

The products, though having identical molecular formulae, differ in colour, solubility and some chemical properties. The two compounds must have different structures and this can be explained if the structure is square coplanar but not if it is tetrahedral. The central Pt^{2+} ion is *dsp*^{2} hybridized with four coordinating ligands, as shown in Fig. 38. In structure (1), the

Structure (1)

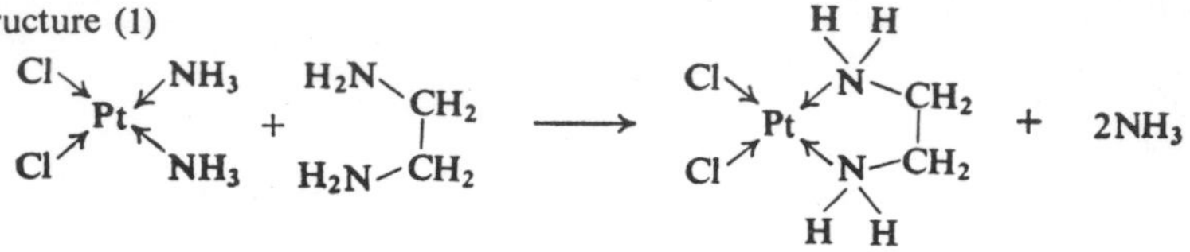

Cl, Cl, NH_3, NH_3 → Pt + $H_2N-CH_2-CH_2-NH_2$ ⟶ Cl, Cl → Pt ← N(H H)–CH_2–CH_2–N(H H) + $2NH_3$

Structure (2)

Cl, H_3N, NH_3, Cl → Pt + $H_2N-CH_2-CH_2-NH_2$ ⟶ NO REACTION

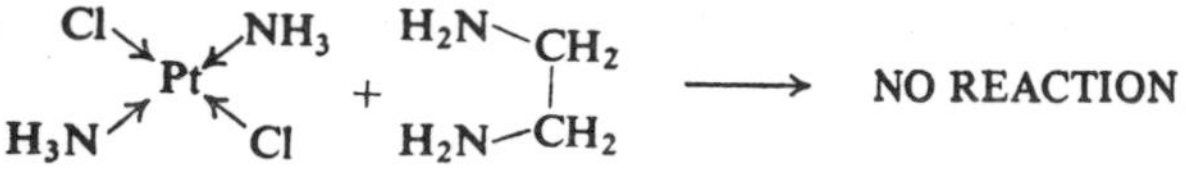

Fig. 38. *Cis-trans* isomers of dichlorodiammineplatinum(II)

ammonia ligands are closer together and can react with ethylenediamine to form a chelate compound, whereas structure (2) cannot.

Structure (1) is *cis* dichlorodiammineplatinum(II) (*cis* from the Latin, meaning 'on the same side') and structure (2) is *trans* dichlorodiammine-platinum(II) (*trans*—meaning 'across').

Optical isomerism

A common cause of optical isomerism among organic compounds is the asymmetry which results from four different groups being covalently bonded, in a tetrahedral arrangement, to a central carbon atom. In a similar manner, a central transition metal ion which is *sp*^{3} hybridized, can be coordinated onto by four different ligands, tetrahedrally arranged, and optical isomerism should result. Only chelates with asymmetrical ligands have, so far, been resolved.

7.6 Stability of complexes

The stability of complexes varies considerably and is influenced by a number of factors.

a. *The nature of the metal*

Metals which form ions of noble gas structure, do not form as many complexes, nor are they as stable as those formed by transition metal ions. Of those which do form, the most stable are those with small ionic ligands, like fluoride in $[AlF_6]^{3-}$, and with ligands which contain oxygen as donor atom, for example EDTA. Smaller metal ions usually form more stable complexes but there are exceptions, as with the magnesium-EDTA complex which is less stable than the calcium counterpart, presumably because such a large ligand cannot find room to coordinate firmly onto the smaller ion.

For transition metals the following sequence of stabilities for bivalent ions has been suggested, Mn^{2+} Fe^{2+} Co^{2+} Ni^{2+} Cu^{2+} Zn^{2+}. When a transition element has different valencies, the complexes of higher valency are nearly always more stable and it is reasonable to assume that the greater the charge, the stronger the power of attraction for electrons.

b. *The nature of the ligand*

For ligands with basic properties, the stronger the base the more stable the complex. Chelate complexes are generally more stable than non-ring complexes, five- or six-membered rings being the most favoured. Ligands with bulky groups only form stable complexes when the coordinated ion is large enough to accommodate the ligand.

When a complex ion is formed, the constituents lose their individual properties. The $[Fe(CN)_6]^{3-}$ complex ion is an example of a stable complex and the equilibrium

$$[Fe(CN)_6]^{3-} \rightleftharpoons Fe^{3+} + 6CN^-$$

lies well to the left. Other complexes, however, which are less stable, may be considerably dissociated in solution, and the presence of both simple ions and complex ions in the solution can be shown. If dissociation of the complex is complete the solution will behave in the same way as a solution of a double salt which shows the properties of all its constituent simple ions. For example aluminium potassium sulphate in solution has properties indicating the presence of K^+, Al^{3+} and SO_4^{2-}.

7.7 Limitations of the explanation of complex formation

The picture of complex formation in terms of ligands coordinating into vacant orbitals provides a useful model, and serves as an introduction to a complicated subject. It does have limitations and, in particular, does not lend itself to quantitative interpretation. Much of the behaviour of complexes can be explained by application of crystal field or ligand field theory. A simplified version of crystal field ideas has already been suggested as being a key to explaining the striking colour changes frequently associated with complex formation and also the magnetic properties of such materials. In ligand field theory, the overlap between ligand and metal orbitals is also taken into account as well as the charge considerations with which the crystal field approach mainly deals. This more detailed treatment is beyond the scope of this book.

8 The Metallic Bond

Elements are frequently described as metallic or non-metallic. What do we mean by this statement? In searching for an answer, it soon becomes clear that it is easier to list a number of characteristics than it is to attempt a formal definition (Table 28).

Table 28. Characteristics of metallic and non-metallic elements

Metallic elements	*Non-metallic elements*
Physical properties	
Solid (except mercury)	Gas, liquid or solid
High melting point	Frequently low melting point
Molten over wide temperature range	Molten over narrow temperature range
High boiling point	Frequently low boiling point
High heat of vaporization	Usually low heat of vaporization
Good conductors of electricity	Poor conductors of electricity (except graphite)
Good conductors of heat	Poor conductors of heat
Opaque	May be transparent or translucent
Lustrous	Not lustrous
High tensile strength	Low tensile strength, if solid
Elastic	Inelastic, if solid
Malleable and ductile	Brittle, if solid
Chemical Properties	
Tend to lose electrons to form positive ions	Tend to gain electrons to form negative ions, or form covalent bonds
Electropositive	Electronegative
Low ionization energy	High electron affinity
Good reducing agents	Good oxidizing agents
Oxides are basic or amphoteric	Oxides are acidic

An element is described as metallic if it has most of the characteristics listed in the left hand column and nearly eighty elements can be classified under this heading. Eighteen elements are best described as non-metallic, and the remaining half dozen lie somewhere in between the two; they are sometimes called *metalloid* elements.

Figure 39 summarizes in the long form of the periodic table, some of the physical properties of the elements and the heavy line encloses the metals. The uncertainty in characterization, as metallic or non-metallic, is at the right-hand side of this block of elements.

8.1 Structure

Metals in the gas phase usually consist of single atoms. The energy needed to effect the change from metal in the solid state, to metal in the vapour

Element	Melting Point (K)	Boiling Point (K)	Heat of Fusion (kJ mol^{-1})	Heat of Vaporization (kJ mol^{-1})	First Ionization Energy (kJ mol^{-1})
H	14	20	0.0586	0.452	1310
He	3	4	–	0.084	1536
Li	453	1600	2.929	147.7	518.8
Be	1550	3040	11.72	314.6	899.6
B	(2300)	–	22.18	313.8	799.1
C	4000	5100	–	719.6	1088
N	63	77	0.3598	2.787	1402
O	54	90	0.2218	3.389	1314
F	53	85	0.7950	3.138	1682
Ne	24	27	0.3347	1.841	2079
Na	371	1170	2.636	99.16	493.7
Mg	923	1380	8.786	127.6	736.4
Al	933	2720	10.46	291.2	577.4
Si	1680	2950	50.63	203.3	786.6
P	317	553	0.6276	12.97	1013
S	392	718	1.213	10.46	1000
Cl	172	238	3.222	10.21	1255
Ar	84	87	1.130	6.276	1519
K	337	1030	2.385	79.08	418.4
Ca	1110	1710	9.205	166.9	589.9
Sc	1810	3000	16.32	305.0	631.8
Ti	1940	3530	15.48	431.0	656.9
V	2170	3720	17.57	460.2	648.5
Cr	2140	2930	14.64	346.9	652.7
Mn	1520	2420	14.64	219.7	715.5
Fe	1810	3270	15.48	251.0	757.3
Co	1760	3170	15.48	382.4	757.3
Ni	1720	3000	17.57	372.0	736.4
Cu	1360	2870	12.97	304.6	744.8
Zn	692	1180	7.364	115.5	907.9
Ga	303	2510	5.607	256.1	577.4
Ge	1210	3100	31.80	334.3	761.5
As	1000	sub.	27.61	32.43	945.6
Se	490	958	5.230	13.81	941.4
Br	266	331	5.272	15.27	1142
Kr	116	121	1.506	9.665	1347
Rb	312	961	2.218	75.73	401.7
Sr	1040	1650	8.786	141.4	548.1
Y	1780	3200	11.30	389.1	636.0
Zr	2120	3850	16.74	502.1	669.4
Nb	2690	3570	26.78	–	652.7
Mo	2880	5830	27.61	535.6	694.5
Tc	2470	–	23.01	502.1	698.7
Ru	2770	5170	25.52	619.2	723.8
Rh	2240	4770	21.76	531.4	740.6
Pd	1820	4250	16.74	376.6	803.3
Ag	1230	2480	11.30	254.0	732.2
Cd	594	1040	6.109	100.0	866.1
In	429	2270	3.264	224.7	556.5
Sn	505	2540	7.196	292.9	707.1
Sb	904	1650	19.83	195.0	832.6
Te	723	1260	17.91	49.79	870.3
I	387	456	7.866	20.84	1008
Xe	161	165	2.050	13.68	1163
Cs	302	963	2.092	68.20	376.6
Ba	987	1910	7.657	149.4	502.1
La	1190	3740	6.276	401.7	539.7
Hf	2490	5670	21.76	648.5	531.4
Ta	3270	5700	28.45	753.1	577.4
W	3680	6200	33.68	774.0	769.9
Re	3450	6170	33.05	636.0	761.5
Os	2970	5770	26.78	677.8	841.0
Ir	2720	5570	27.61	636.0	887.0
Pt	2040	4800	21.76	510.4	866.1
Au	1340	3240	12.68	342.3	891.2
Hg	235	630	2.343	58.16	1008
Tl	576	1730	4.268	162.3	589.9
Pb	600	2000	5.104	177.4	715.5
Bi	544	1830	10.88	178.7	774.0
Po	527	–	–	121.3	–
At	(575)	–	–	33.47	–
Rn	202	211	3.347	16.40	1038
Fr	(300)	–	–	–	–
Ra	973	–	10.04	114.6	–
Ac	1320	–	–	–	–
Ce	1070	3740	5.021	397.5	665.3
Pr	1210	3400	6.694	330.5	556.5
Nd	1290	3300	7.113	288.7	606.7
Pm	(1300)	–	–	–	556.5
Sm	1340	2170	8.786	192.5	539.7
Eu	1100	1710	9.205	175.7	548.1
Gd	1580	3270	15.48	301.2	594.1
Tb	1630	3070	16.32	292.9	648.5
Dy	1680	2870	17.15	280.3	659.9
Ho	1730	2870	17.15	280.3	–
Er	1770	3170	17.15	280.3	–
Tm	1830	2000	18.41	246.9	–
Yb	1100	1700	7.531	159.0	598.3
Lu	1920	3600	19.25	376.6	481.2
Th	2020	4120	19.25	–	–
Pa	(1500)	–	–	543.9	–
U	1400	4090	11.30	460.2	–
Np	910	–	–	–	–
Pu	913	3500	–	–	–
Am					
Cm					
Bk					
Cf					
Es					
Fm					
Md					
No					
Lw					

Fig. 30 Physical properties of the elements

phase, is high. Apparently metal atoms in the solid are held together by attractive forces of considerable size. We have seen how electron transfer and the resulting formation of ionic lattices, or electron sharing to form covalent or coordinate bonds, provides a model for the formation of compounds and accounts for the binding power which holds the atoms or ions of a substance together. How can we account for the forces between identical atoms in a metal? What is the nature of the *metallic bond*? What model will help us to understand?

X-ray interference patterns produced from metal foil show that in the majority of metals the atoms can be regarded as spheres, regularly packed, in one of three ways;

i. hexagonal close packed,

ii. face-centred cubic close packed,

iii. body-centred cubic packed.

a. *Hexagonal close packed spheres*

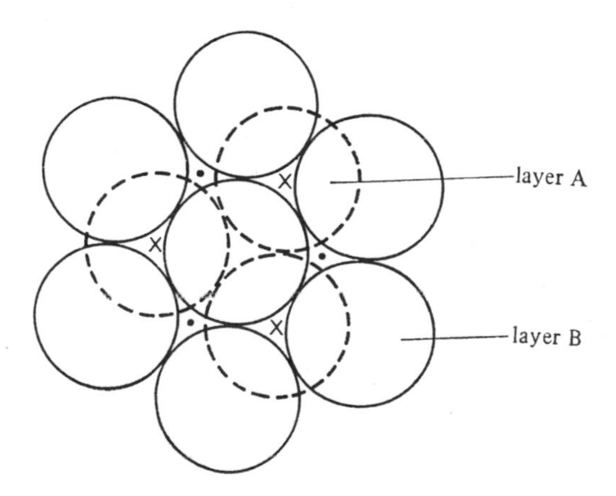

Fig. 40. Close packing of spheres, plan view

In layer B, each sphere is in contact with six other spheres (Figs. 40 and 41). Layer A is identical, but is placed on layer B so that the spheres of layer A, occupy the hollows in layer B, marked by a cross. The third layer is directly below layer A. This arrangement is called ABABAB, and each sphere is in contact with twelve other spheres (Fig. 41). The coordination number is twelve and the arrangement is said to be close packed, 74 per cent of the space being occupied.

b. *Face-centred cubic close packed spheres*

Layers A and B are as for hexagonal close packed, but the third layer is placed so that the centres of the spheres are below the hollows in layer B marked with a dot (Fig. 40). This arrangement is ABCABC and each atom

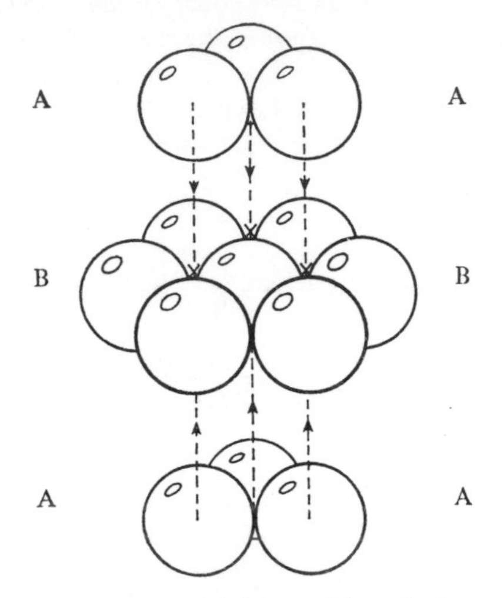

Fig. 41. Hexagonal close packing of spheres

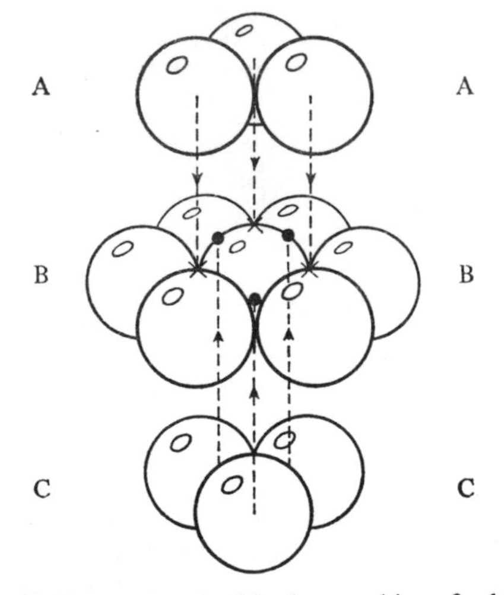

Fig. 42. Face-centred cubic close packing of spheres

is surrounded by twelve neighbours (Fig. 42). Again the coordination number is twelve, the arrangement is close packed, and 74 per cent of the space is occupied.

c. *Body-centred cubic packed spheres*

The basic unit is a cube with one sphere at the centre in contact with eight others at the corners (Fig. 43). Here the arrangement is more open. The coordination number is eight, only 68 per cent of the space is occupied and the arrangement is not described as close packed.

Fig. 43. Body-centred cubic packing of spheres

Table 29 shows the type of packing encountered in metal structures.

Table 29

Hexagonal close packed	*Face-centred cubic close packed*	*Body-centred cubic packed*
Beryllium	Aluminium	Barium
Cadmium	Calcium	Caesium
Calcium	Chromium	Calcium
Lithium	Cobalt	Chromium
Magnesium	Copper	Cobalt
Strontium	Gold	Iron
Titanium	Iron	Lithium
Zinc	Lead	Manganese
	Lithium	Molybdenum
	Manganese	Potassium
	Nickel	Rubidium
	Platinum	Sodium
	Scandium	Strontium
	Strontium	Tungsten
	Silver	Vanadium

Several metals, for example calcium, can exist in more than one form; they are said to be *polymorphic*. The different crystalline forms are stable over different temperature ranges, the transition temperature being the temperature at which the different forms can exist in equilibrium. Such behaviour is an example of enantiotropic allotropy encountered earlier. The structural reorganization necessary to switch from one form to another is not great, such changes could be expected to occur fairly easily.

8.2 The metallic bond

Metal atoms then, are arranged in a regular array, with every atom in close contact with several neighbours. It is this close packed structure which accounts for the high density of most metals, although the atomic volume of individual atoms is an important factor.

Metals have unoccupied outer orbitals and these overlap from atom to atom, and the valency electrons are free to move through the delocalized orbitals which result. Since each metal atom loses, to some degree, direct control over its valency electrons it is helpful to imagine the structure of a metal as a lattice of metal ions, through which permeates a sea or gas of electrons. The force of attraction between the ions and the electrons overcomes the repulsion forces between particles of like charge, and holds the structure together.

This simple electron gas model of the structure of a metal is adequate to explain in a qualitative way the properties of metals, but as will be indicated later, a rather more sophisticated treatment is required if a quantitative approach is adopted.

8.3 Melting points and boiling points

With few exceptions, metals melt at temperatures above 800 K and remain liquid over a wide temperature range, about 1000 K.

Imagining the metal crystal to consist of close packed metal ions in a sea of delocalized electrons, it follows that considerable energies will be needed to separate the atoms. When sufficient energy is supplied to collapse the solid, to form a melt, there is only a small increase in volume and the opportunities for orbital overlap are not much reduced. Much energy is still required to effect further separation of metal atoms, and the molten phase persists over a wide temperature range. The general similarity in energy content of the solid and liquid phases is reflected by the relatively low heats of fusion, usually not more than one or two kilojoules per mole.

The final step of separating the atoms to a position beyond that where the attraction for each other has any significance, has to overcome the delocalized effect completely, and the high energy requirement results in high heats of vaporization, (80 to 400 kJ mol^{-1}), and high boiling points.

8.4 Electrical conduction

When a potential difference is applied to a metal, the delocalized electrons nearer to the positive connection can flow towards it. Other nearby electrons move to take their place and a general drift of electrons throughout the structure occurs. For every electron leaving the metal at the positive terminal, an electron enters via the negative terminal.

The increased movement of electrons increases their overall kinetic energy and some of this is transmitted to the metal atoms. The average kinetic energy of the system increases and so the temperature rises. Raising the temperature of a metal increases the vibrational motion of the metal atoms and this interferes with the free flow of electrons through the matrix, increasing the electric resistance.

When a quantitative treatment of the conductance of metals is attempted, it is found that the simple electron gas model begins to break down. This approach is also inadequate in that it does not account satisfactorily for the existence of semi-conductors such as germanium; a rather more sophisticated model refines the above simple outline and postulates the existence of *energy bands* within the structure of a metal.

In a free atom, electrons are arranged in a few discrete energy levels outlined in Fig. 1. In a metal lattice, however, only the innermost electrons are arranged in this way. The outer energy levels on neighbouring atoms interact to produce energy bands within which there is a continuous distribution of energies. The small number of valency electrons associated with each metal atom thus has available many vacant orbitals through which they can move, and the electrons no longer belong to a particular atom, but to all the atoms in the vicinity. When an electric field is applied to the metal, the electrons can therefore take up small amounts of energy and move in the direction of the field. On this model, the electric conductance of a metal will be inversely related to the number of valency electrons, since the larger the number of valency electrons the smaller the number of orbital vacancies available. This prediction is borne out in practice, and it is usually found that divalent metals have a greater electric resistance than monovalent metals.

For non-metallic elements, the number of valency electrons is such that the levels in the energy bands are fully occupied and hence thermal or electrical excitation does not readily lead to an uptake of energy since there are no empty levels of suitable energy available.

In semi-conductors, although many of the energy bands are fully occupied, suggesting insulating properties, there is available a band of only slightly higher energy which is unoccupied and into which electrons can readily pass when thermally excited. Promotions brought about in this way means that unpaired electrons and vacant levels are now available in both energy bands, and the electrons can therefore move under the influence of an electric field. It follows that the ability of a substance such as germanium to act as a conductor varies directly with temperature; increased temperature causing the promotion of more electrons. For a metal on the other hand conductance falls with increased temperature.

One further important difference in behaviour between metals and semi-conductors is that the conducting properties of the former generally fall in the presence of traces of impurity; for a semi-conductor, the reverse seems to be true, and it is assumed that the impurity atoms furnish additional energy levels, and make transitions between energy bands easier.

8.5 Conduction of heat

When one part of a metal is heated, the metal atoms vibrate with greater vigour, and some of this motion is passed on to the delocalized electrons in the neighbourhood. Whereas the thermal agitation of metal atoms or ions can only be passed on to other atoms in direct contact, the delocalized electrons can accelerate the process by carrying energy through the lattice. Non-metals, without delocalized electrons to act as energy carriers, transmit heat much more slowly and are bad conductors at ordinary temperatures.

8.6 Strength

The tensile strength of metals is a measure of the attractive force between the metal ions and the sea of electrons associated with them. If a perfect array is imagined as extending throughout the metal, the metal will be soft since the planes of atoms can slide over each other readily. Cubic close packed structures can undergo slip in any one of four directions, and as a consequence are malleable and ductile. Hexagonal close packed structures have only one direction of slip resulting in reduced malleability and ductility. If, as is likely, there are a number of vacancies, or holes, in the structure, bending is facilitated since atoms can successively slip into the vacancies and a row of atoms slide gradually through the structure. When a metal in which the atoms are imagined to be perfectly aligned is sharply bent or hammered, the extensive regular array breaks up into smaller pockets of crystallinity. Dislocations can no longer run through the structure so readily and the metal becomes harder.

Elasticity can be accounted for by supposing that atoms move slightly out of position under stress, returning to their original places when the stress is removed. If the stress is too great the atoms are displaced from the original position and the metal is permanently bent.

8.7 Lustre

Light striking the surface of a crystalline metal is absorbed by the loosely bonded electrons near to the surface causing them to move about more quickly. The rapidly moving electrons emit radiant energy and the consequence of absorption and radiation of energy is that the surface 'reflects' light; it is lustrous.

8.8 Impurities

The presence of even small quantities of an impurity decreases the ease with which slip can occur and reduces malleability and increases hardness. This can be a disadvantage, as when traces of arsenic present in copper make the drawing of wire impossible, and reduce the electricalcon ductance by interrupting the movement of electrons through the lattice.

On the other hand, careful control of additives can produce alloys with advantages over the pure metal. Alloys are solid solutions, and in simple cases a proportion of metal atoms of one kind, are replaced by atoms of another in the lattice. For this to be possible the atomic radii must not differ by more than 15 per cent and the constituent elements must be similar in properties and in valency. In other alloys there is evidence of covalent bond formation within the structure.

8.9 Interstitial Compounds

The carbides, nitrides and borides of transition metals are close packed structures, with the smaller non-metallic atoms occupying the holes between the larger metal atoms.

There is little change in the metallic bonding, as instanced by the electrical conductance of the 'compounds', but the presence of the non-metal atoms renders the structure very hard, of exceptionally high melting point and boiling point and chemically inert.

9 Knotty Problems

9.1 Paramagnetism of oxygen

Oxygen is paramagnetic and when placed in a magnetic field, the molecules experience a weak attractive force. Paramagnetism is associated with structures containing at least one unpaired electron, and the structure for oxygen given in Chapter 2, has none.

According to the molecular orbital theory only the two 1*s* electrons on each atom belong separately to the two oxygen atoms. The remaining six electrons from each atom, are accommodated in new molecular orbitals as follows: the four 2*s* electrons are paired, one pair occupying a *bonding orbital* and the other pair an *anti-bonding orbital*, four of the 2*p* electrons are paired in each of two bonding orbitals. This leaves two 2*p* electrons which go separately into two anti-bonding orbitals, and these are supposed to be the electrons of unpaired spin responsible for the paramagnetism.

The simple treatment of molecular structure, which has been adopted throughout the book, generally succeeds in giving a satisfactory explanation of behaviour. The failure to account for paramagnetism need not undermine confidence in the simple interpretation, but does serve as a reminder that a model represents but does not mirror reality.

9.2 Ozone

When an electric discharge is passed through oxygen, ozone is formed. It is unstable, having a higher energy content than oxygen, and its formation is spontaneously reversible. Molecular weight determination shows ozone to be triatomic and its structure is usually represented (Fig. 44a) with a double bond joining two oxygen atoms, one of which is coordinated on to a third.

(a) 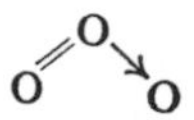 O=O→O

(b) 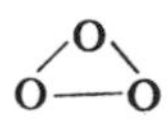 O—O—O (ring)

Fig. 44. Ozone

It is impossible to visualize this structure on the valence bond theory without assuming that the oxygen atom accepting the coordinate bond, pairs up the two $2p^1$ electrons to vacate a 2*p* orbital for occupation by the coordinate bond pair of electrons.

The molecular orbital theory does offer an interpretation of this structure but it is not dealt with here.

It is tempting to use the valence-bond picture to postulate a ring structure for ozone (Fig. 44b). This could be accounted for in terms of atomic orbitals, but calculation from bond energies shows it to be even less stable than ozone actually is, and X-ray diffraction data discounts the ring arrangement and proves that the bonds in ozone are equal in length and inclined at an angle.

9.3 Carbon dioxide

Assuming sp^3 hybridization for carbon, there seems to be little difficulty in devising a structure for carbon dioxide (Fig. 45). The carbon atom forms two sp^3–p bonds with each of the two oxygen atoms, pairing and sharing all unpaired electrons. Unfortunately, this picture does not account for the measured bond lengths. The C═O length is usually 0·121 nm, but in carbon

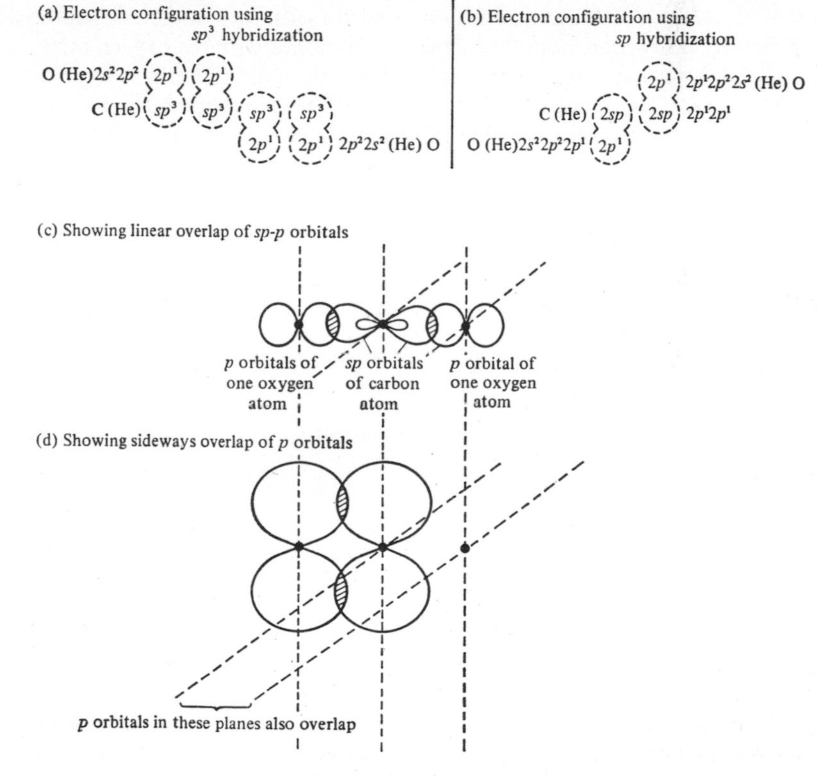

Fig. 45. Carbon dioxide

dioxide the carbon and oxygen atoms are 0·113 nm apart. One possibility, to account for this discrepancy, is to suppose that s–p hybrid orbitals are formed.

The electron configuration for carbon is $(He)2s^2 2p^1 2p^1$, and promotion of one $2s$ electron to the vacant $2p$ orbital gives $(He)2s^1 2p^1 2p^1 2p^1$. Hybridization of the $2s$ orbital with one of the $2p$ orbitals results in $(He)(2sp)^1(2sp)^1$-$2p^1 2p^1$. Each of the sp orbitals overlaps with a p orbital on each oxygen atom (Fig. 45). The two remaining $2p$ orbitals of the carbon atom, mutually at right angles to each other and to the axis of the molecule, overlap sideways with the $2p$ orbitals on each of the two oxygen atoms, giving a model which helps to explain the shortening of bond length.

9.4 Carbon monoxide

Carbon monoxide can be represented as a straight forward coordinate structure, but is included here for comparison with carbon dioxide. In the structure, the carbon atom forms two $2p$–$2p$ bonds with the oxygen atom, and the latter uses an unshared pair of electrons to coordinate into the vacant $2p$ orbital of the carbon atom (Fig. 46). It is interesting to try to write a structure for carbon monoxide in which the carbon atom forms two sp hybrid orbitals and to speculate whether this structure is a better model than the one shown in Fig. 46.

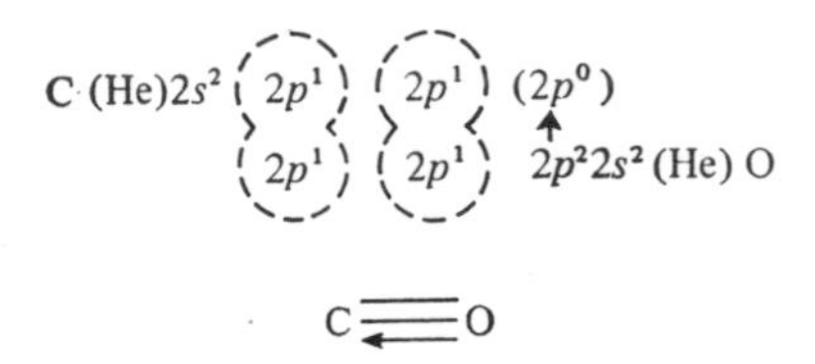

Fig. 46. Carbon monoxide

9.5 Silicon dioxide

The properties of silicon dioxide are in marked contrast to those of carbon dioxide, despite the fact that carbon and silicon are the first two elements of group IV. The high melting point, 1983 K, suggests that forces of higher magnitude hold the structure together.

In silica, the molecule SiO_2 does not exist. Instead there is a giant molecular network of silicon–oxygen bonds, tetrahedrally arranged as in diamond. This can be explained by assuming the formation of sp^3 hybrid orbitals for silicon. The electron configuration of silicon is $(Ne)3s^2 3p^1 3p^1$, and promotion of a $3s$ electron to the vacant $3p$ orbital gives, $(Ne)3s^1 3p^1 3p^1 3p^1$, which hybridizes to $(Ne)(3sp^3)^1(3sp^3)^1(3sp^3)^1(3sp^3)^1$. Each of the four sp^3 hybrid orbitals overlaps with a p orbital of an oxygen atom to give a three-dimensional network (Fig. 47).

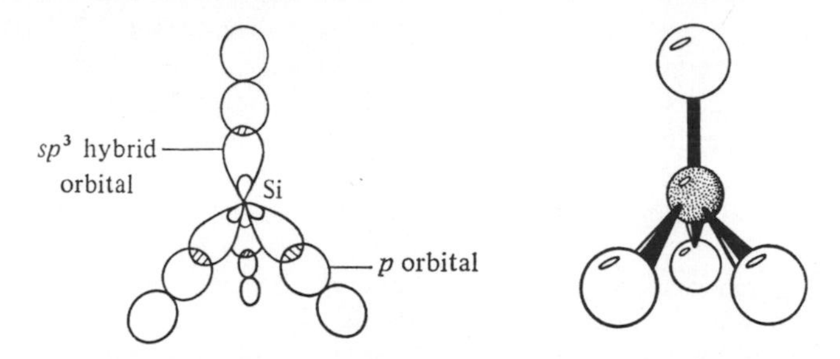

(a) Basic silicon-oxygen tetrahedron

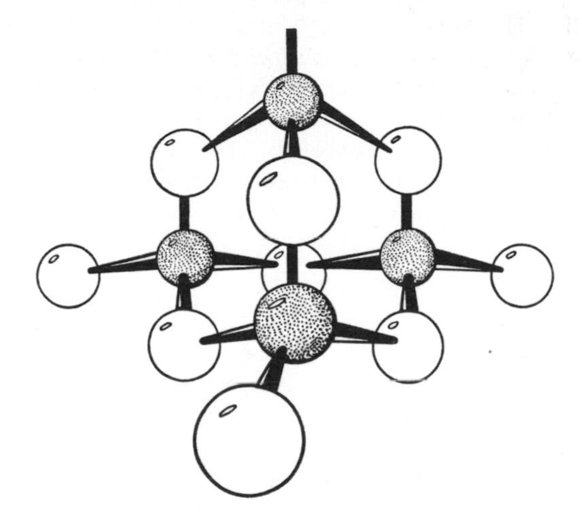

(b) Crystal structure (silicon atoms shaded)

Fig. 47. Silicon dioxide

The high energy needed to break the silicon–oxygen bonds explains the high melting point and resistance to chemical attack of silicon dioxide.

9.6 The oxides of nitrogen

The three familiar compounds, dinitrogen monoxide, nitrogen monoxide and nitrogen dioxide present formidable problems.

Dinitrogen monoxide

It is known that the molecule is linear and consists of a central nitrogen atom joined to a nitrogen atom and an oxygen atom. The configurations of nitrogen and oxygen are respectively $(He)2s^22p^12p^12p^1$ and $(He)2s^22p^22p^12p^1$, and normal overlap between the two $2p$ orbitals of one nitrogen atom with the two $2p$ orbitals of the oxygen atom satisfies the pairing of the oxygen. The pairing of the remaining $2p$ electron on the central nitrogen atom with a $2p$ electron on the other nitrogen atom, gives complete pairing for the central nitrogen atom but leaves the other one with two unpaired electrons. This is an unlikely structure and invalidated by the fact that dinitrogen monoxide is not paramagnetic. Similar problems beset other attempts to find a structure for this molecule.

Two structures which are often given for dinitrogen monoxide are $\overset{-}{N}=\overset{+}{N}=O$ and $N\equiv\overset{+}{N}-\overset{-}{O}$ and it is impossible to represent them in terms of atomic orbitals. Dinitrogen monoxide has a very small dipole moment, 0·54 C m, and this is not in keeping with either of these charged structures. It is interesting that the dipoles operate in opposite directions along the axis of the molecule, and it is tempting to speculate in terms of mixtures of the separate species. However, evidence for their existence has not been obtained. Pauling discusses such a structure in terms of the resonance theory, but this is not considered here.

Nitrogen monoxide

Nitrogen monoxide is paramagnetic, so at least one unpaired electron is to be expected in the structure. The valence bond treatment cannot go beyond suggesting a structure with two *p–p* bonds joining the nitrogen atom to the oxygen atom, the unpaired electron remaining in a $2p$ orbital of the nitrogen atom (Fig. 48). It has been suggested that some sort of three electron bond is formed.

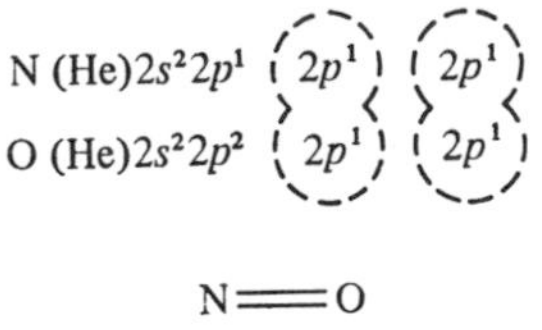

N $(He)2s^22p^1$ $2p^1$ $2p^1$

O $(He)2s^22p^2$ $2p^1$ $2p^1$

N$=$O

Fig. 48. Nitrogen monoxide

It is interesting that the dimer, N_2O_2, is diamagnetic in the liquid state, suggesting that the odd electron has been paired and shared between the two NO molecules.

Nitrogen dioxide

Nitrogen dioxide is also paramagnetic and a possible structure shows (Fig. 49) an unshared *p* electron remaining on the oxygen atom. Again, this structure is not adequate, since it is found from infra-red studies of NO_2 that the two N—O bonds are identical in length (0.119 nm). The difficulty is partially resolved by dealing with this structure in terms of resonance

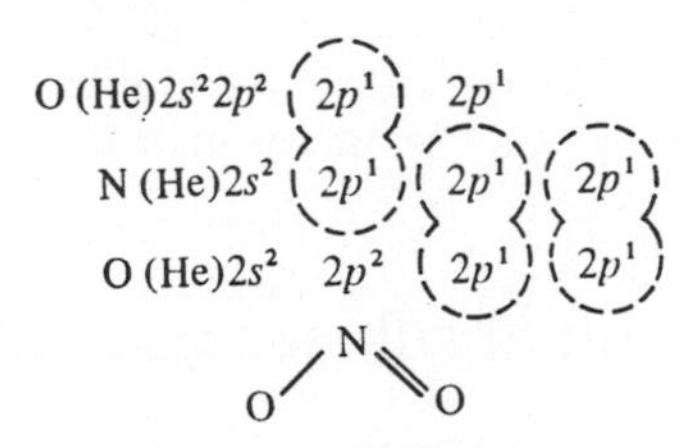

Fig. 49. Nitrogen dioxide

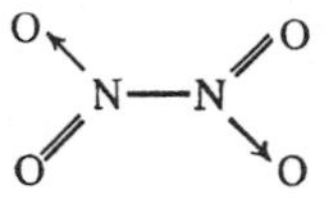

Fig. 50. Dinitrogen tetroxide

theory. Dinitrogen tetroxide is diamagnetic, and evidence from diffraction studies suggests that the monomers are joined by a bond between the nitrogen atoms. A possible structure is given in Fig. 50, though this too is an oversimplification. The N—N bond in the dimer is very long and therefore weak; suggestive of the ease with which N_2O_4 dissociates.

10 Summary

An attempt is made in this chapter, to group substances. The groupings are based upon the principal types of electron rearrangement which have been discussed earlier.

The coordinate bond does not feature separately in the summary, because like a covalent bond, it is a shared pair of electrons. It differs only in that the shared pair was originally supplied by one atom or group of atoms.

The generalizations are sweeping and are guide lines rather than precise descriptions. It must also be remembered that substances cannot always be sharply differentiated, but may be intermediate in type.

10.1 Classification of substances under bond type

Monatomic (A)

Argon	Krypton	Radon
Helium	Neon	Xenon

Molecular non-polar (A_x)

Bromine	Hydrogen	Oxygen
Chlorine	Iodine	Phosphorus (white)
Fluorine	Nitrogen	Sulphur

Molecular non-polar (A–B, *with cancelling dipoles*)

Acetylene	Carbon tetrachloride	Mercury(II) chloride
Aluminium chloride	Complexes	Methane
Beryllium chloride	Ethylene	Phosphorus pentoxide
Boron trichloride	Hydrazine	Phosphorus trioxide
Carbon dioxide	Mercury(I) chloride	Tin(IV) chloride
Carbon disulphide		

Molecular polar (A–B)

Ammonia	Hydrogen sulphide	Sulphur dichloride
Antimony trichloride	Hydroxylamine	Sulphur dioxide
Bismuth trichloride	Hypochlorous acid	Sulphur trioxide
Carbon monoxide	Iodic acid	Sulphuric acid
Carbonyl chloride	Iodine monochloride	Sulphurous acid
Chloric acid	Nitric acid	Sulphuryl chloride
Chlorous acid	Orthophosphoric acid	Thionyl chloride
Complexes	Perchloric acid	Tin(II) chloride
Hydrogen cyanide	Phosphine	Water
Hydrogen halides	Phosphorus trichloride	
Hydrogen peroxide	Sulphamic acid	

Giant molecular (A_x)

Carbon (diamond)	Carbon (graphite)	Phosphorus (red)

Giant molecular ($(A{-}B)_x$)

Copper(II) chloride	Silica	Silicon carbide

Giant ionic lattice ($(A^+B^-)_x$)

Most metallic salts, oxides and hydroxides. Ammonium salts.
Some degree of covalent bonding, within the lattice, is not uncommon.

Metallic

Metals

10.2 Relationship between bond type and behaviour

	Monatomic	*Molecular non-polar*	*Molecular polar*	*Giant molecular*	*Giant ionic lattice*	*Metallic*
Grouping of atoms	A	A_x or A–B with cancelling dipoles	A–B	A_x or $(A\text{–}B)_x$	$(A^+B^-)_x$	A_x
Type of bond in structure	None	Electron pair	Electron pair	Electron pair	Ionic	Metallic
Intermolecular forces	van der Waals'	van der Waals'	van der Waals' and dipole–dipole	—	—	—
Physical state	Gas	Gas, liquid, solid	Gas, liquid, solid	Solid	Solid	Solid
Melting point/ Boiling point	Very low	Low	Fairly low	Very high	High	Usually high
Heat of fusion	Low	Fairly low	Fairly low	Very high	High	Intermediate
Heat of vaporization	Low	Fairly low	Fairly low	Very high	High	High
Electrical conductance	Zero	Zero	Very low	Zero, except graphite	Zero for solid, high for melts	High
Electrical conductance of aqueous solution	Zero	Zero, except when reaction with water produces ions	Usually high, as ions formed by reaction with water	—	High	—
Solubility in water	Low	Low, but higher if reaction with water produces ions	High	Very low	Varies but often soluble	Very low
Solubility in covalent solvents	High	High	Fairly high	Very low	Low	Very low

Strength	—	Very poor	Very poor	Frequently very hard	Hard and brittle	Malleable and ductile
Thermal stability, resistance to bond rupture by heat	—	Usually high, but depends upon degree of overlap, so greater for smaller atoms	Usually high, but depends upon degree of overlap	Very high	Very high	High
Reacivity under ordinary conditions	Low	Variable, depends upon degree of overlap and environment	Variable depends upon degree of overlap and environment	Low	Variable, but not generally reactive	Variable, but generally reactive

Book List

ADDISON, W. E. *Structural Principles in Inorganic Compounds*, Longmans 1963

AYLWARD, G. H. and FINDLAY, T. J. V. *Chemical Data Book*, second edition, Wiley 1966

AYNSLEY, E. E. and DODD, R. E. *General and Inorganic Chemistry*, Hutchinson 1963

BELL, C. F. and LOTT, K. A. K. *Modern Approach to Inorganic Chemistry*, Butterworth 1963

BOND, G. C. *Principles of Catalysis*, Royal Institute of Chemistry Monographs for Teachers No 7

BREY, W. S. *Physical Methods for Determining Molecular Geometry*, Chapman & Hall 1966

BUTTLE, J. W., DANIELS, D. J. and BECKETT, P. J. *Chemistry: A Unified Approach*, Butterworth 1966

CHEMICAL BOND APPROACH PROJECT, *Chemical Systems*, McGraw-Hill 1964

CHEMICAL EDUCATION MATERIAL STUDY, *Chemistry: An Experimental Science*, Freeman 1963

CHOPPIN, G. R. and JAFFE, B. *Chemistry: Science of Matter, Energy, and Change*, Silver Burdett 1965

DEWAR, M. J. S. *An Introduction to Modern Chemistry*, Athlone Press of the University of London 1965

GOULD, E. S. *Inorganic Reactions and Structures* Holt, Rinehart & Winston 1963

GREENWOOD, N. N. *Principles of Atomic Orbitals*, Royal Institute of Chemistry Monographs for Teachers No 8

HARVEY, K. B. and PORTER, G. B. *Introduction to Physical Inorganic Chemistry*, Addison Wesley 1963

JAMES, R. *X-ray Crystallography*, Methuen 1961

JENNINGS, K. R. *Molecular Structure*, Chapman & Hall 1964

LARSEN, E. W. *Transition Elements*, Benjamin 1965

LEE, J. D. *Concise Inorganic Chemistry*, Van Nostrand 1964

LYNCH, P. F. *Orbitals and Chemical Bonding*, Longmans 1966

MCGLASHAN, M. L. *Physico-chemical quantities and Units*, Royal Institute of Chemistry Monographs for Teachers No 15

MOODY, B. J. *Comparative Inorganic Chemistry*, Arnold, 1965

NEBERGALL, W. H. and SCHMIDT, F. C. *General Chemistry*, Heath 1959

PAULING, L. *The Nature of the Chemical Bond*, Oxford University Press 1963

PAULING L. and HAYWARD, R. *The Architecture of Molecules*, Freeman 1965

RYSCHKEWITSCH, G. E. *Chemical Bonding and the Geometry of Molecules*, Chapman & Hall 1965

SANDERSON, R. T. *Principles of Chemistry*, Wiley 1963

SANDERSON, R. T. *Teaching Chemistry with Models*, Van Nostrand 1962

SPEAKMAN, J. C. *Molecules*, McGraw-Hill 1966

SPICE, J. E. *Chemical Binding and Structure*, Pergamon Press 1964

SISLER, H. H. *Electronic Structure, Properties, and the Periodic Law*, Chapman & Hall 1965

STEELE, D. *The Chemistry of the Metallic Elements*, Pergamon Press 1966

WILSON, J. G. and NEWALL, A. B. *General and Inorganic Chemistry*, Cambridge University Press 1966

Index